GROW FOOD FOR FREE

HUW RICHARDS

CONTENTS

INTRODUCTION

Imagine eating seasonal produce you've grown and harvested with your own hands as part of every meal. Imagine having unlimited access to your own greengrocers' 365 days a year. Imagine growing food that costs next to nothing.

These are not fantasies but realistic aims and objectives. And I can assure you they are achievable. How can I be so sure? Over the last 12 months, I challenged myself to grow food for free. Now I want to pass on what I learned on that journey, including strategies to deal with any issues that may arise. I also offer different growing options, depending on whether your space is small or large, in the country or in the city, so you can choose what works best for you. Everything in this book has been tried and tested, and I hope that the information it contains gives you the confidence to grow food that tastes just as good as you imagined.

WHY COST IS NO OBSTACLE

Over the years, I have heard a whole host of reasons why people are reluctant to start growing their own food, and one of the most common is that it is too expensive. I find it deeply saddening that this misconception persists in society and that, as a result, so many people are not only put off, but are also missing out on a fantastic opportunity. Fired up by a strong desire to set the record straight, I've written this book with the intention of passing on all the information you need to grow food for free in a single location.

Growing food for free involves basic gardening techniques, such as sowing and planting, but making a success of it requires you to look at things a little differently. Changing your mindset and seeing the value in what others might regard as trash is key. A broken pallet from a building site can be split and made into a compost bin, a box of empty milk cartons from a café will provide vital water storage, and a pile of cardboard outside a shop entrance is the perfect material for suppressing weeds around your crops.

WHAT DOES "FREE" MEAN?

Before we get started on the growing journey, I feel it is important to set down my definition of "free". To me, "free" means obtaining something without any money changing hands. A simple example would be exchanging seeds of a vegetable you don't like for those of one you'd much rather grow. And in this book, "free" doesn't mean "for no effort" because – and let's get real here – it is impossible to grow food without putting in time and energy. Even foraging for wild food requires effort, as does buying food from a shop. Of course, money may make some tasks easier and save time, but there are ways to avoid spending if you are resourceful and open to opportunity.

TAKING SMALL STEPS

Giving up too soon is the one thing that will prevent you enjoying free food in abundance. To prevent this happening and ensure successful outcomes, start slowly so you don't feel overwhelmed. Always split bigger projects into smaller, attainable goals and write down these smaller tasks in the form of a "to-do" list. By carrying out lots of smaller tasks when time permits and crossing them off the list, you begin to think of yourself as highly efficient and productive. Having made lots of small steps, you will feel you're constantly making great progress, and that will keep you motivated and on track. In the following pages, I'll explain what I mean by "bartering" and how this useful skill is key to growing food for free.

THE POWER OF BARTER

Now that I've explained my definition of "free", I'd like
to introduce you to what I believe is the best alternative
out there to money – the skill of bartering.

The *Oxford English Dictionary* defines bartering as the "exchange of goods or services for other goods or services without using money". What this straight-to-the point, objective definition can't describe is how powerful bartering can be when used as a framework for social interaction. Before monetary systems were developed, people traded by bartering goods and services, and each exchange involved engaging with another person or group, which is what I love about it.

HOW AND WHAT TO BARTER
A key aspect of bartering is the lack of guidelines; there is no "one size fits all" system. Bartering, or swapping one useful thing for another, is a tool that anyone can use, and it has the key advantage of flexibility. You can offer different "bartering items", depending on what is of most interest or value to the person you are trading with, or it can be as simple as doing someone a favour by taking away what they (but not you) consider "waste".

You don't even have to offer a physical item. Your time, your skills, and your knowledge in a particular area are also a form of currency. If, for example, someone is offering fence posts, but they are more interested in your guitar-playing skills than the items you've put up for exchange, you could offer them a guitar lesson instead.

CONNECTING WITH OTHERS
It's exciting when you take money out of the equation because every exchange is unique. The individuals or groups involved must negotiate the best agreement for all, and there can be quite a thrill to this exchange. If you have never bartered before, you will be pleasantly surprised at the simplicity of the process, the enthusiasm of those engaging in it, and the amazing range of things on offer. Bartering is one of those things which can seem scary at first, but you very quickly ease into it.

This book features many examples of useful things you can obtain through bartering, such as composting materials and seed packets. And for me, one of the key benefits is making new connections with people. Over time you will find that the people you barter with will start offering items without you even asking, or put you in contact with other like-minded people. You then have the option of creating a small bartering group, which in turn makes it much easier to set up tool shares (see p17). You can do as little or as much bartering as you like, but I find it a much more rewarding form of exchange than handing over money.

BARTERING IS A TOOL THAT
ANYONE CAN USE, AND IT
HAS THE KEY ADVANTAGE
OF FLEXIBILITY

I'm always collecting odd items from neighbours (*top left*). In exchange I offer them produce I have grown (*top right, bottom right*) or seeds that I've collected from my crops (*bottom left*).

SET UP YOUR VEG PATCH

FINDING AND EVALUATING AN AREA TO
GROW FOOD IN, AND SOURCING THE
TOOLS YOU NEED TO GET STARTED

WHAT YOUR SPACE NEEDS

The first part of any gardening project should be to evaluate your (or even find an) outdoor space. I use the acronym SWAGA when assessing an area's potential for growing food. The letters stand for Size, Water, Aspect, Ground, and Access.

SIZE

You don't need as much space as you might think to grow a decent amount of produce. For example, an area of 1 x 2m (3 x 6½ft) is enough to supply a family of four with salad leaves from mid-spring to late autumn. To put that into context, a standard parking space is just under 3 x 4m (10 x 13ft). Incidentally, this is a great-sized plot for new fruit and vegetable growers to start with!

WATER

This precious commodity is vital for crops in spring and summer, but is heavy and awkward to transport. A roof is the best source of free rainwater, which can be collected and stored during the wetter months, and used when the sun is shining. Greywater (waste water from sinks, showers, and baths) is perfectly fine to use on the garden and is a fantastic resource in periods of very low rainfall. If you must use mains water, get permission from the person who pays for the supply.

QUICK CHECKLIST

Use this list to ensure your chosen site has the basic features needed to successfully grow food.

- Space for a compost bin, water storage, and the crops you want to grow
- Water access – from a roof or greywater
- At least four hours of direct sunlight in spring and summer, and shelter from strong winds
- Ground covered by lawn or tiles, or a terrace
- Access without going through a building (if possible)

ASPECT

The aspect determines how much direct sunlight an area receives each day. South-facing plots get the most sun, so allow for the widest range of plants; east-facing come second on the list. Even spaces that only get four hours of direct sunlight a day are suitable for most crops. Plant leafy greens in areas of reduced sunlight – they don't mind shade and will still be productive. Shelter from strong winds will help prevent plants getting damaged or even blown over.

GROUND

A space covered by lawn offers the most options for growing crops, as you can dig a bed, or place raised beds or containers on top. On hard surfaces, such as patios or terraces, raised beds and containers are your only options. If you are renting your home, speak to your landlord before establishing your setup, and certainly before you start digging. Don't forget that you can take containers with you when you move.

ACCESS

In the excitement of acquiring a space in which to grow food, access is often overlooked and can cause problems later. Restricted access is common for small outside spaces, such as enclosed courtyards. If your space is accessible only via your house, consider whether you are prepared to carry large and/or heavy materials through your indoor living space? If you are planning on using a large area, you will want to have wheelbarrow access.

When evaluating the potential of a new space, I like to do a few quick drawings of how I could set up the area for growing.

TIP

If you're planning on growing a variety of crops, consider using two locations – the closest to your house dedicated to labour-intensive crops, such as salads, and the furthest for lower-maintenance ones.

TOOLS TO SHARE OR BORROW

Less is more (and definitely cheaper) when it comes
to garden tools – so much can be achieved using just your hands.
However, there are some tools that are well worth having.

Your hands are vital for so many tasks that require dexterity, such as sowing seeds, pricking out and potting on seedlings, tying in peas and climbing beans, or checking soil moisture. I rarely wear gloves because it is important for me to be in direct contact with the soil to feel its texture and moisture levels.

KEY TOOLS
A few tools – a spade, rake, fork, trowel, hammer, knife, secateurs, drill, and saw – are all you need to grow your own food. Although buying these new would be expensive, there are cheaper options (*see opposite*). Wheelbarrows are useful, but you can use large buckets instead, especially if you only have a small space.

Your hands are the only tool you need for a range of gardening tasks – including making holes when you transplant young plants.

a. Spade
For cutting turf, digging holes for transplanting directly into the ground, and filling up large containers with compost or soil.

b. Rake
To sweep up leaves or grass clippings and to prepare ground for planting. Use the end of the handle to create shallow trenches for sowing seed.

c. Fork
For turning compost, harvesting deep-rooted crops and digging up cuttings to plant out elsewhere.

d. Trowel
To fill containers and modules with compost, for transplanting, and for removing large weeds.

e. Hammer
Very useful for building compost bins and raised beds and mashing up brassica stems before composting.

f. Knife
To cut string, to harvest produce, and for many other small tasks.

g. Secateurs
For pruning perennials, for harvesting, and for preparing hardwood cuttings.

h. Drill
To securely screw together wood when building your own compost bins and raised beds.

i. Saw
For building structures, such as raised beds, and other garden projects.

a

b

c

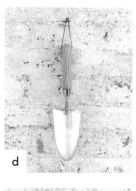

d

e

f

g

i

h

SOURCING TOOLS FOR FREE

There are two quick and easy ways to source the tools you don't have, or exchange existing tools for others. The first option is to organize a tool share. Get together with friends or neighbours and ask everyone to make a list of their tools. Once you know what is available, you can arrange a mutually convenient time to borrow the tool you need. As long as everyone respects each other's property and pays for breakages, the system works well.

Online or community swap shops are another excellent source of gardening tools. Every household has unloved items that someone else could put to good use so why not exchange these for garden tools? No money needs to change hands, so I would define this as free. Used tools that once belonged to elderly gardeners may also be offered on online freecycle or swap sites. These are often good quality and well cared for.

USE SOMEBODY ELSE'S GARDEN

If you don't have your own outdoor space, don't fret! Many people have neither the time nor the energy to look after their gardens, and would be delighted if you could breathe new life into their outdoor space in return for some fresh food.

Before you take on someone else's space, I would advise you to respect the owner's wishes above everything else. If they want to keep the lawn intact, don't dig up even a small area. Clear communication between yourself and the owner is vital, so always talk through ideas and projects with them to avoid any misunderstandings and, at worst, losing the space.

HANDY CHECKLIST

Make sure your expectations are realistic by running through the SWAGA checklist (see pp14–15), and asking yourself the following questions:

- Is the space within walking/cycling distance of my home?
- Do I get along well with the garden owner?
- Are there many restrictions on the space?
- Is the sole access through the owner's house?
- How regularly can I visit the space?

BE PREPARED TO SAY NO

Listen to your instincts and if you are at all unsure, don't take on that particular garden. Both you and the owner need to feel completely comfortable with the arrangement, as well as one another.

There are so many people looking for someone to take care of their gardens – including one of my neighbours (*above*) and this lady (*left*), who was my London bed-and-breakfast host on a trip to visit my publisher. Sadly, I couldn't help because I live in Wales.

OBTAINING SPACE TO GROW

Below, I've listed some suggestions of how and where to start your search for growing space.

Local social media groups

The best groups are the "swap shop" types where people trade, sell, or look for recommendations. Using social media is the easiest and most effective way of reaching a large proportion of your local community. People can tag those offering space or put you in touch with others who aren't in the group.

Flyers

Decide how far you are willing to travel or target a specific street and deliver handwritten notes. The most effective are short, polite, and to the point.

Local gardening groups

People often contact local gardening groups to offer their garden to someone who has the time and energy to look after it. Look online, in community centres, or in local newsletters for your nearest group.

MAKING THE ARRANGEMENT WORK

The key to success in any arrangement between you and the owner of the garden is to make sure you both benefit. If you are actively searching for land, it is always important to offer a share of your harvest as an incentive. Everyone likes the prospect of eating home-grown healthy food, grown without any effort on their part, and so no money needs to be exchanged.

Establish from the outset whether the owner would like a proportion of the harvest, a supply of their favourite vegetable, or a small box of produce on a regular basis.

URBAN GARDENING

Community gardening schemes are probably your best option if you are looking for a space in a town or city. Many will allow you to grow whatever you'd like with permission from the organisation running it (whether the local council or a business that wants to support local sustainability projects). For more information about this, look at movements like Incredible Edible (see p222).

Guerrilla gardening is a term used to describe growing food on land that you don't own, rent, or have the right to cultivate, and is often done on abandoned sites and in secret (see the resources on p222 for some great examples). It is a controversial practice, and I personally wouldn't recommend it because you could lose all your hard work if the area is reclaimed or the owners destroy your plants because they don't want people on the site.

THE KEY TO THE SUCCESS OF
ANY ARRANGEMENT IS THAT
ALL PARTIES BENEFIT.

GROW FOOD ON A PATIO, TERRACE, OR BALCONY

Often oriented to make the most of the sun, patios and terraces can prove a fantastic space in which to cultivate crops. Use large containers to grow a wide range of food, including climbing beans, potatoes, and salad.

Coming up with strategies to pack in as much as you can to make the most of these limited spaces can be exciting. If you do enough planning up front and use the planting methods outlined below, your patio or terrace could be more productive than a small plot.

INTERCROPPING

This technique involves growing slow- and fast-maturing crops in the same space. As they grow at different rates, they will both have sufficient space – the quickest-maturing plants will be harvested and removed before the slower-growing ones reach full size. It's more efficient, as you can plant the slower-growing crops immediately, rather than having to wait.

Native Americans had an intercropping method called "Three Sisters", and used it to grow squash, sweetcorn, and climbing beans in the same bed. The tall sweetcorn stems provide a structure for the beans to climb, while the spreading squash foliage acts as a living mulch to suppress weeds and retain soil moisture.

SUCCESSION PLANTING

Plan ahead and have one crop ready for planting when you've harvested another. For example, start spinach seedlings on your windowsill as your runner beans begin to flower. When the beans have finished cropping, remove the plants and plant out your spinach seedlings. This saves time between harvests, as plants are established and ready to transplant at the point at which there is an area available for them in your growing space (see p42 for more information).

VERTICAL GROWING

Growing crops on or up walls or supports is a way of increasing yield when space on the ground is limited.

You can attach gutters to walls or fences to grow salad leaves, train trailing courgettes up stakes, and grow peas or beans up trellises. Try standing up pallets against walls, lining the gaps with hessian or old compost bags, and then filling them with compost to grow herbs and strawberries. South-facing walls absorb heat and make perfect backdrops for growing sun-loving plants such as tomatoes, cucumbers, and peppers in containers.

SMALL-SPACE COMPOSTING

Patios and terraces often lack space for making compost. Also, if your terrace has a wooden surface, the moisture from a compost bin could rot the timber. The best place for a compost bin in this situation is on a paving slab or similar hard surface. Alternatively, you could line the bottom of your compost bin with plastic from old compost bags to retain any moisture and protect the wood beneath.

BALCONIES

Balconies usually have more complicated access than patios and terraces, and may also have a weight restriction. One key benefit of a balcony is that it will be slug-free (unless you accidentally bring them in on plants) and there will space for a container or two as well as wall areas for vertical growing.

Best crops for balconies
- Salad leaves
- Climbing beans
- Potatoes
- Tomatoes
- Courgettes
- Peas

Sunny terraces provide a great opportunity to grow food (*above and above left*). Use space-saving methods, such as planting up an upright pallet (*left*), to make the most of the space.

BEST PERENNIAL FRUIT AND VEGETABLES FOR CONTAINERS

- All herbs (see pp92–97)
- Jerusalem artichokes (see pp100–101)
- Blackberries and hybrid berries (see pp102–105)
- Strawberries (see pp106–109)
- Currants and gooseberries (see pp110–115)

ESTABLISH A WATER SUPPLY

Water and compost are the gardener's two most precious resources. Mains water is reliable yet expensive, but rainwater and greywater are both free and can provide fantastic alternative sources of water.

Aim to establish water storage and a water supply before the growing season is in full swing, so you don't run into shortages at the most crucial points in the year.

RAINWATER

Although rainwater is free, the challenge is capturing and storing it. Rainwater that runs off roofs into gutters and drainpipes can easily be diverted for collection in large water butts.

For a free alternative rainwater diverter, try cutting the gutter downpipe on your house off around 1.5m (5ft) from the ground. When rain is forecast, put a large bucket, dustbin, or even a wheelie bin underneath to gather water. When you aren't collecting water in a container, put large stones around the sides of the drainage hole to funnel water water into the drain below.

Shed roofs, especially those made of corrugated metal, are another good source of free water. Line buckets along the side wall, just outside the roof overhang to capture rainwater.

If you have space, you can create your own rainwater collection system. Prop a large piece of board or metal sheet against something big and flat at an angle. You can then use an old piece of guttering (easily sourced from online swap shops or free recycling schemes) to collect the rainwater run-off and divert it into a container.

STORING RAINWATER

Once you start capturing rainwater, store as much of it as possible, so that you always have a supply during periods of low rainfall or drought. Large milk cartons, car screenwash containers, dustbins, beer kegs, wine barrels, and even old galvanized baths can all be put to good use. Use whatever you have to hand and can source for free – the more containers you fill, the more water you have for growing crops.

GREYWATER

Water from baths and sinks that would otherwise disappear down the drain is much better used on your garden. Whenever you do some washing up,

When rain is forecast, line up buckets against the wall of a shed or garage without a gutter – the water collected may prove to be an invaluable supply during periods of dry weather.

TIP

If the container attached to your downpipe is full, use it to fill up any other empty water containers you have. That way, you will always be at maximum capacity.

for example, use a bowl and empty it into a bucket or watering can to use on seeds, seedlings, and growing plants. Make sure that any soap or washing up liquid you use is made from 100 per cent natural ingredients because it will be kinder to both soil and plants. To make collection easier, you could divert the sink waste pipe into a bucket. Greywater won't store well, so I try to use it within a couple of days.

SOIL AND WATER RETENTION

It is a fundamental truth that the healthier your soil is the less you need to water it. Follow no-dig gardening methods (see p68) and add plenty of organic matter – in the form of home-made compost or well-rotted manure – to significantly increase your soil's capacity to hold and retain water. Even when there has been no rain in my garden for a period of two to three weeks, I only need to water seedlings, plants in containers, and salad leaves. Other crops cope well without water because the soil structure is so healthy.

Place a large container under a cut-off downpipe when rain is forecast to collect water from the roof of your house (*left*). I use old milk cartons to store extra rainwater (*above*).

WORK WITH YOUR CLIMATE

Trying to grow food without understanding the prevailing weather conditions is like baking bread without knowing how high or low to set the oven. Fortunately, you can avoid potential disasters by learning to protect your plants from frost.

Climate and weather are interchangeable terms to many people, but they do mean different things. Climate refers to the long-term weather conditions prevailing in a particular region, whereas weather describes the day-to-day conditions such as sun, rain, wind, temperature, and so on in a particular area. Global climate change has resulted in unpredictable weather patterns and extreme weather events, so when growing your own crops, it's always best to be prepared – especially for frosts and droughts.

BENEFITS AND DRAWBACKS

Different climates always have their advantages and disadvantages. For example, I garden in a cool, damp climate, so my crops need less watering than they would in a hot, dry climate, and brassicas and root vegetables do really well for me. That said, I have daydreamed about having enough sun and warmth to grow tomatoes and aubergines outdoors.

To get the most from your growing area, it makes sense to focus your energy on crops that are best suited to your location. Researching your local climate online and matching food crops to those conditions is time well spent, and, crucially, you will come to understand the frost patterns in your area.

MAKE THE MOST OF THE SUN

The position of your beds and containers will have a big impact on their productivity. As mentioned earlier, south- and east-facing boundaries get the most sun. Ideally, these boundaries will have a wall or fence to provide a warm microclimate for your crops, absorbing sun during the day, and radiating heat at night. If you do have an ideal border, prioritize sun-loving crops, such as peppers and tomatoes, in these locations.

FROSTS

A frost occurs when the outside temperature falls below 0°C (32°F) and causes moisture to turn into ice crystals. Some vegetables cannot tolerate frost, which causes their living cells to suffer irreversible and even fatal damage. Yet once you know your average first (before winter) and last (before summer) frost dates, the time between them represents a large window of growing opportunity. See p222 for links to websites that will tell you the relevant frost dates for your area.

HARDINESS

Annual vegetables are organized into three groups according to their hardiness (their ability to tolerate cold). Hardy vegetables can withstand temperatures down to -8°C (18°F) (and sometimes colder), half-hardy vegetables can tolerate very light frosts for a few hours, and tender vegetables cannot survive a frost at all. Hardiness is provided for every vegetable in the Annuals chapter (pp116–157) for easy reference.

Give tender vegetables, such as tomatoes, a head start by sowing seeds indoors in late winter. You can then transfer the young plants outside a few weeks after the last frost and get a longer cropping period. I typically begin transplanting half-hardy annuals two weeks after the last average frost and three to four weeks after it for tender plants.

MY FOUR-DEGREE RULE

When a freak frost strikes, it can take you completely by surprise and you risk losing all your carefully nurtured tender plants. My strategy is to monitor several weather forecasts and if any of them indicate the night-time temperature might fall to 4°C (39°F) or

a

b

c

below, I always take precautions. Weather forecasts are never 100 per cent accurate, so I chose this temperature to be on the safe side, and I always stick to the rule. Even if the temperature only dips to two or three degrees, heat-loving plants will still benefit from the protection.

PROTECT YOUR CROPS FROM FROST

You may have already transplanted tender or half-hardy vegetables in spring when you notice that temperatures are forecast to drop. If this happens, it's vital to protect your crops as soon as possible. Here are three quick and easy methods, using material you are likely to have at hand, for when my four-degree rule kicks in:

a. Bed sheets

Place equal height sticks in the ground around the plants and drape over an old bed sheet to provide insulation. Two sheets would be even better. Do this before sunset to help retain some of the day's warmth.

b. Leaves or straw

Have a bag of leaves, straw, or hay at the ready to cover young plants and seedlings overnight. Collect the material the following morning and return it to the bag, so that you can use it again if necessary.

c. Cardboard box

Empty cardboard boxes turned upside down act as small, insulated "rooms" and work well for individual plants, such as squash. Always weigh down the box with stones if it is windy to stop it from blowing away.

"FAILING TO PREPARE IS PREPARING TO FAIL" IS A SAYING THAT SEEMS PARTICULARLY RELEVANT TO GARDENERS IN FROST-PRONE REGIONS.

EXTREME WEATHER

Putting strategies in place for dealing with floods, droughts, heatwaves, hail, or snow will help to minimize damage, or can even prevent the loss of an entire crop. Perennials tend to be far more resilient than annuals and will cope better with extremes, although those growing in containers won't survive without water in periods of drought. The following suggestions offer maximum crop protection for minimum effort in a range of extreme weather conditions.

HAIL

Plants at highest risk: Seedlings and salad leaves
Crop issue: Damage to salad leaves

A sudden hailstorm can easily catch you unaware. Ideally, protect young seedlings and salad leaves from damage by throwing a light blanket (or similar covering) over them before a forecast storm starts. If you're caught out by a sudden hailstorm, just provide protection as soon as you can. Salads grow very quickly, so any damaged leaves on these plants will soon be replenished.

SNOW

Plants at highest risk: Tender plants and brassicas
Crop issues: Collapse

A cardboard box or the stick-and-sheet method (see p25) both offer protection from snow for salad leaves, but when using the stick and sheet method, always remember to scrape snow off the sheet so it doesn't sag and squash the plants beneath. Hardy vegetables should cope without any protection, but always knock any snow off tall brassicas.

WIND

Plants at highest risk: Legumes
Crop issues: Collapse

Grow the tallest crops in the most sheltered spots to minimize wind disruption, and push pea and bean supports deep into the ground to create a strong structure. These should provide sufficient stability for the plants' growing period from spring to early autumn when storms are less likely, but are still a risk.

DROUGHT

Plants at highest risk: Seedlings, young plants, salads
Crop issue: Harvest failure, bolting

In periods of unusually low rainfall, prioritize watering young plants and seedlings – they can't store as much water because their root systems are not yet fully developed. Drought can stress some plants, such as salad leaves, into bolting (producing flowers and seed prematurely), which ruins the flavour. Water your crops regularly during droughts to help prevent this. Sowing new salad crops every two weeks will keep your supply going if you lose a crop.

PLANTS HAVE DIFFERENT ABILITIES TO WITHSTAND WEATHER ISSUES, SO GROW A RANGE OF CROPS TO MINIMIZE DISAPPOINTMENT.

Check forecasts and protect crops from extreme weather, such as hail (*top left*), strong wind (*top right*), snow (*bottom right*), and droughts that cause bolting (*bottom left*).

CHOOSE YOUR SETUP

Watching a patch of bare earth fill up with an abundance of food is hugely motivating. Below, I discuss the key features of containers, raised beds, or growing directly into the ground.

CONTAINERS

Growing in containers is great fun, easy to do, and can provide fantastic crops. Although plants in pots need more water than those in raised beds or the ground, this is the quickest way to establish a corner of your garden or your balcony to grow food. I always suggest starting to grow food in a few containers, as it doesn't require as much time or effort as the other methods given here. Try creating your own containers, like I have with this tyre (*below*, see pp32–33).

Three top performers:
- potatoes
- salad leaves
- herbs

RAISED BEDS

I grow almost all of my vegetables in wooden raised beds (*below*). They need more time and investment upfront compared to container growing, but you can grow a far greater range of plants and they will need much less watering! Set up a raised bed on stone, concrete, or turf, but not wooden decks or patios, which will rot. Raised beds drain well, even in periods of heavy rainfall, so plant roots don't sit in wet soil, starved of oxygen and prone to disease or rot.

Three top performers:
- strawberries
- root vegetables
- garlic and onions

IN THE GROUND

The key benefit of growing directly in the ground is that most of the growing medium (the topsoil) is available to you for no extra effort. The depth of soil also makes this method more suited to growing deep-rooting plants and perennials, such as rhubarb (*right*), although it's wise to check for buried cables or drainage pipes in urban areas. If you suffer from back problems, a raised bed may be easier to work in.

Three top performers:
- rhubarb
- brassicas
- climbing beans

THE BEST OPTION?

The ideal scenario is to have a mix of all three methods, because plants do have preferences. In reality, the best option for you will depend on your situation. Throughout the book I will give what I believe to be the most effective way of growing a particular crop to ensure you get the most out of your space without spending money.

COMPARING THE METHODS

	CONTAINERS	RAISED BEDS	IN THE GROUND
Easy to move	Y	N	N
Suitable for small spaces	Y	Y (for smaller beds)	N
Gardens with no soil	Y	Y	N
Low watering	N	Y	Y
Low compost	N	Y	N
Low weeding	Y	Y*	Y*
Minimal tools	Y	N	Y
Easy setup	Y	N	Y
Suitable for perennials	Some	Y	Y
Suitable for rented homes	Y	Get permission first	Get permission first
Suitable for patios and balconies	Y	N	N

* Using the mulch method (see p124)

CREATE YOUR OWN CONTAINERS

You will need containers for raising seedlings, growing mature plants, and propagating cuttings. Wherever you live, you'll find containers available for free. I hope my suggestions might inspire you to come up with your own ideas!

FOR SEEDLINGS

Small containers are best for sowing seeds or potting up soft cuttings, such as herbs. Propagating hardwood cuttings, such as blackcurrants, however, is best done in large containers, raised beds, or in the ground.

a. Cardboard loo- and kitchen-roll tubes

These tubes have two key advantages: they are easily sourced and completely biodegradable. When seedlings are ready to transplant, simply plant the entire container and its contents. Beetroot, chard, and brassica seedlings all grow well in carboard tubes.

Cardboard tubes are perfect for crops such as peas and beans, which have long roots. When you pick the tubes up for transplanting, some compost inevitably falls out. One solution is to make four 2.5cm (1in) cuts around one end of the tube, and then fold in the resulting tabs to create a base.

b. Newspaper pots

Biodegradable newspaper pots can be made with a paper-pot maker, and transplanted with their contents in the same way as cardboard tubes (*above*). If you don't have a pot maker, wrap a piece of paper around a small, sealed tin, so there is a good amount of overhang over one end, and fold the overhanging paper flat over that end. Take a second, smaller tin, and use it to press down on the folded paper to firm in place. Remove the finished pot from the tin.

c. Tin cans and yogurt pots

These upcycled containers can be used for three to five years before being given a final rinse and put in the recycling. Squash and chard seeds will do well in yogurt pots, but be sure to poke a few holes in the base for drainage before you use them.

d. Guttering

I've been hooked on this wonderful method for a while, using it primarily for starting off pea seedlings. Sliding them off the gutter into a trench in a raised bed has to be one of the most satisfying gardening tasks! Cut the gutter down into smaller sections to suit the size of your outdoor space. Also, make sure you raise guttering off the ground when you have peas or beans growing in it. This will prevent mice eating your crops. You can easily source guttering from skips, or from friends and neighbours who are having roofs or sheds repaired or refurbished.

e. Egg boxes

Cardboard egg boxes make perfect biodegradable modules for starting off seedlings and are readily available – save your own or ask friends and family. Brassicas, salads, and peas will all germinate and grow well in egg boxes, as will annual herbs. When it comes to transplanting the seedlings, simply tear off the individual cups and plant out the whole thing.

a

b

c

d

e

FOR MATURE PLANTS

MAKE YOUR OWN CONTAINER

Everyday household items or discarded objects are free, and easily converted into containers for growing food. Here I've used a tyre to create a container for some herbs, but you can adapt these steps to fit the objects you have. Check p36 to ensure your chosen object is deep enough for the crops you want to grow.

1. Source an old tyre for free from either a farm or garage (you will be doing them a favour because commercial collection and disposal costs money).

2. Collect some empty compost bags from friends or neighbours. Cut them to size, poke large holes in them, and use them to line the inside of your tyre. Use two bags if one isn't enough.

3. What you use to fill the bottom of your container will depend on what you are growing. Here I have used small stones, as the herbs I'm planting like well-drained soil. If you want to grow crops that like moist soil, add twigs, prunings, leaves, and grass clippings.

4. Fill the tyre with a 50/50 mix of soil and compost and let this settle for a few days before giving it a final top up. I mixed in some old sand from a sandpit at this stage so that the soil drained better for my herbs.

5. You now have the ideal container for growing a range of crops – I've planted herbs in mine, but you could try growing salad leaves, beans, or strawberries.

OTHER OBJECTS TO USE AS CONTAINERS
Buckets

Used buckets or catering-sized tubs with drainage holes allow you to grow a whole crop of lettuce or strawberries. Get them for free (as my friend does) from a nearby café, or ask neighbours who buy buckets of fat balls for birds. Local builders often throw out leaky buckets, which are perfect for growing plants in when filled with compost.

Shopping baskets

Discarded supermarket shopping baskets contain large holes, but you can use an old compost bag or piece of sacking as a liner. The handles also make it very portable. You might find rusty baskets in alleyways

or hedgerows. If you can't tell where they came from, by all means help yourself; however it's best to return them if you can!

Cardboard boxes

Shallow cardboard boxes lined with an old compost bag make excellent containers, especially for fast-growing crops, such as radish and salad leaves. Once you've harvested the crop, tear up the box and compost it, saving the plastic for the next container.

Beer kegs

Repurposed plastic beer kegs make fantastic containers for runner beans and potatoes. I found a couple of local landlords who were happy to give away their empty beer kegs, but you can always offer seasonal produce in exchange to get the ball rolling.

Paint cans

Water-based paint cans make good containers for climbing beans and salad leaves. Those with handles are perfect for when you want to move salad leaves out of the hot sun. Wash the cans thoroughly before use and remember to make drainage holes in the base.

This cardboard box has been lined with an old compost bag in the same way as the tyre planter (see steps, *opposite*), and used to grow lettuce.

Milk cartons

Large milk cartons with their tops cut off and drainage holes in the base work well for smaller plants, such as salads and annual herbs. Alternatively, and this works particularly well on a balcony, cut off the base instead and securely tie cartons upside-down to a rail, poking holes in the caps for drainage.

YOU CAN EVEN TURN
OLD BATHS INTO RAISED BEDS
(IF YOU CAN FIND ONE!)

Raised beds can be made from a variety
of materials, such as pallet collars (*top
left*), logs (*top right*), reclaimed bricks
(*above*), and old tractor tyres (*right*).

CHEAP RAISED-BED SOLUTIONS

You can make a low-cost frame for a raised bed using a variety of easily salvaged materials or weave your own from pliable willow. Another option is to forget permanent edging altogether and make a simple heaped bed.

Below I list a variety of materials you can use to make make raised beds. My raised beds are no larger than 1.2 x 3m (4 x 10ft). This width allows me to reach the middle from both sides and the bed isn't so long that I am tempted to hop over rather than walk around it!

Breeze blocks
Building sites and builders' merchants are good places to get hold of imperfect or chipped blocks. They are heavy, stable, and often hollow, providing space to plant herbs or flowers. Building the bed is straightforward and you can extend it by adding a few extra blocks – much easier than creating a second bed.

Large stones or reclaimed bricks
For a very rustic-looking raised bed, make sides of large stones or bricks sourced from demolished walls in rural areas – with permission. There will be gaps between the stones, so keep on top of any weeds growing through. With more stones and a little creativity, try making dry-stone-wall sides.

Logs
Felling an unstable tree offers the perfect opportunity to create a beautiful raised bed from logs. Ask friends or neighbours, as they may have tree work scheduled and no use for the wood. If you can get it sawn into regular lengths, it will be easier to arrange the wood into raised-bed sides. Once the logs start to break down, use them for *hügelkultur* (see p38).

Woven beds
Approach rural landowners for permission to collect willow and hazel branches during winter. Weave these together and with a little practice you can create amazing raised beds, even round ones! The branches will need to be replaced over time, so maintain good contact with your source to ensure a future supply.

Tractor tyres
Just one huge tractor tyre creates an instant raised bed for growing produce. Line the inside with old compost bags (making drainage holes in them first) to prevent any tyre chemicals from leaching into the soil.

Pallet collars
This is by far the easiest DIY raised bed of them all. Pallet collars come packed flat. When you open them out, you get an instant raised-bed frame.

OTHER POSSIBILITIES
I've suggested a few ideas for DIY raised beds to get you started – there are so many more possibilities out there for transforming or repurposing other materials and objects. Reclaimed wood may be the most easily sourced material, but if you can only find chemically treated wood, use old compost bags to line the sides.

CREATE A RAISED BED FOR FREE

A great way to create a raised bed without needing to buy any materials is to upcycle an old pallet. Everything is repurposed – the wooden planks, the blocks, and even the nails.

Once used, pallets are often seen as industrial waste and a fire hazard. This means that, wherever you live, you won't have to go far to find someone willing to give them away. Only use pallets marked with an HT symbol, which indicates that they have been heat-treated, and not treated with chemicals that could leach into the soil.

WHERE TO SOURCE PALLETS

Pallets are widely used wherever goods are transported, so consider starting your hunt in the following places:

- Farms and agricultural merchants
- Warehouses
- Sports clubs
- Council waste facilities
- Online swap sites, such as GumTree and Freecycle

Leave your pallets outside in the rain, and plan to make your raised bed soon after. The rain will soak the wood and make the pallets easier to dismantle.

HOW MANY PALLETS?

I found that I only needed one pallet to make a 1.2 × 1.2m (4 × 4ft) raised bed that was two planks deep. Although this is quite shallow, it's still deep enough to grow plenty of different crops, so don't worry if you only have a few pallets but want to make multiple beds. If you have a more plentiful supply, try making deeper beds so that you can grow a wider range of fruit and veg (*see below*).

One pallet can make a raised bed measuring 1.2 x 1.2m (4 x 4ft) and two planks deep.

HOW DEEP SHOULD MY RAISED BED BE?

 1 PLANK Lettuce, spinach, chard, radish, beetroot

 2 PLANKS Smaller carrots, strawberries, perennial herbs, beans

 3 PLANKS Brassicas, potatoes, carrots, leeks, rhubarb, Jerusalem artichokes

 4 PLANKS Blackberries, gooseberries

1

2

3

4

HOW TO BUILD

These are instructions for how to build a raised bed that is two planks deep, but they can be adapted to suit the depth you want. Gather the required tools and materials for the size of bed you want to build before you start (*see below*).

1. Lay your pallet on the ground (*see photo*) and break it down into the parts listed below – I use a pry bar and hammer for this. Splitting a pallet may seem tricky, but you will quickly get the hang of it. Organize eight of the planks into four pairs to form the four sides of your raised bed. There will be two planks left over, which will be cut into shorter "cross" planks to hold the pairs together.

2. To make your first cross plank, lay one of the two spare planks across one of the pairs (*see photo*). Measure and cut the plank so that it is the width of the pair of side planks. Repeat this process until you have eight short cross planks.

3. Position one cross plank 15cm (6in) from either end of one of the four pairs of side planks. Nail into place with the short nails to hold the pair together securely. Repeat for each of the four sides of your raised bed.

4. Line up one of the pallet blocks with the end of one of the newly created sides, and nail it in place with two long nails. Do the same with the next side, so it is at a 90-degree angle to the first, and nail it to the block to create a corner. Repeat for each corner to complete your raised bed. If you have enough, add extra blocks to the second layer to strengthen the sides.

TOOLS AND MATERIALS

You will need a pry bar, hammer, and saw, plus:

For a two-plank-deep bed
- one pallet broken up into 10 equal-length planks
- 32 salvaged short nails
- 8 salvaged long nails
- 4 pallet blocks

For each extra plank in depth you will need
- 6 more planks
- 16 more short nails
- 8 more long nails
- 4 more blocks

FILL YOUR RAISED BED

Once you have created a raised bed and have compost or soil to hand, filling it is straightforward. The methods below use a variety of organic materials to ensure that you have a nutrient-rich growing medium for your crops.

When considering how to fill your raised bed, don't worry if you're short of compost or soil (see pp52–53). Neither compost nor soil are required for either the composting method or the in-fill method.

THE STANDARD METHOD

When I build new raised beds, I usually fill them with a 50/50 mix of compost (or very well-rotted farmyard manure) and topsoil. If you don't have much compost, fill the bed with soil to 10cm (4in) below the top, then finish with a 5cm (2in) layer of compost. Use liquid feed to promote plant growth (see pp72–75), and add a mulch of compost each autumn.

If you don't have much compost, you can fill the bed to the final 10cm (4in) with topsoil, and then fill to the top with a shallow layer of compost.

THE COMPOSTING METHOD

Using this method, the bed acts as a compost bin. You can't start to plant anything until the second year, but you will have a very nutrient-rich growing medium when you do!

- **Year 1** Add materials that are easily broken down, such as grass clippings and kitchen scraps, but avoid twigs and woodchip. Build up the layers, mixing everything thoroughly at the start of every month and cover the bed with cardboard over winter.
- **Year 2** If you are keen to get growing in the second year, create gaps in the cardboard in late spring and plant courgettes (see p147). They love the nutrition, but won't take up much root space. Expect huge yields.
- **Year 3** The bed will be ready to plant with your choice of crops.

THE IN-FILL METHOD

This method takes inspiration from *hügelkultur* (*see opposite*) and is a great way of making use of the topsoil that already exists beneath your lawn. You'll need to have some lawn space on which to put your raised bed, and enough bulky organic material to fill the hole (I use woodchip, see p70).

- Mark the final location of the raised bed. Slice off the turf from this area of lawn, and then excavate about 15cm (6in) of topsoil, keeping both the turf and soil.
- Add the turf to the base of your hole, then position your raised bed around the hole. Fill the hole with the organic material so the surfaces is about 5–10cm (2–4in) up the sides of the bed – it will sink over time.
- Finally, cover the organic material with the soil you dug out from beneath the turf. Top up your raised bed with your homemade compost when you have some.

THE HÜGELKULTUR METHOD

With this method you don't need a large amount of soil and compost because it makes use of different layers of organic matter. *Hügelkultur* is most suitable for deeper raised beds and relies on materials breaking down over time to provide a slow, sustained release of nutrients to crops growing in the bed. It also helps to retain moisture in periods of dry weather.

1. Spread out newspaper over the bottom of the bed to prevent any grass or weeds growing through. Then fill the bottom of the bed with large branches that are 3–5cm (1¼–2in) in diamater. You can also use logs as long as they are less than one-third of the height of the bed in diameter.

2. Add a thick layer of woodchip, then a layer of straw, grass clippings, used farm-animal bedding, and/or autumn leaves (in whatever proportions you like) so the bed is slightly more than two-thirds full. These materials are light, so they will shrink down.

3. Fill the bed to the top with a 50/50 mix of topsoil and compost. Allow the bed to settle for at least a week (ideally two or three weeks), before topping up with more compost or, at the very least, a 5cm (2in) layer of compost at the top. You can now start planting your crops.

DID YOU KNOW?

Hügelkultur literally means "hill" or "mound culture" in German. It generally describes a steep-sided raised bed, nurtured with rotting wood and planted with crops.

CREATE A BED IN THE GROUND

By turning an area of lawn into a bed you can grow an abundance of annuals, such as beans, squash, potatoes, brassicas, and beetroot. You just need a spade's depth of soil, good preparation, and enthusiasm.

THINGS TO CONSIDER

Soil saturation
Plants can suffer when sitting in wet soil that drains slowly after periods of prolonged rainfall. On the plus side, they won't need as much water in hot, dry spells. Adding quantities of homemade compost will improve both drainage and water retention.

Grass invasion
Spend some time edging and maintaining your bed to stop grass infiltration. One low-cost solution is to bury planks from dismantled pallets around the perimeter, or cover the area around the bed with cardboard and then woodchip.

Gardening at ground level
To state the obvious, beds in the ground are low down. If you suffer from back problems, raised beds and containers might prove a better option.

Slugs and snails
Beds without edging, especially those surrounded by grass, allow slugs and snails easy access. See pp160–163 for tips on how to protect your crops.

Expansion
If you want to extend your range of crops, growing in the ground opens up opportunities. A system of beds with enough space between them to get a wheelbarrow through will allow you to make the most of the space.

1

2

PREPARING THE AREA

Stripping turf off the surface is usually advised if you don't want lots of weeds, but you can create a bed without digging by covering the grass with a thick sheet mulch. You will need a rake, cardboard, enough compost or well-rotted manure to create a 5cm (2in) mound, and enough woodchip to cover the compost. If you don't have a good compost supply, wait until the following year. Start in autumn so the bed has time to settle and is ready for planting in spring.

1. Cut the chosen area of lawn as short as possible, and cover it with at least three layers of cardboard to prevent the grass growing through.

2. Spread a 5–7cm (2–2¾in) layer of compost over the cardboard. If you have a plentiful supply of compost, add up to a 15cm (6in) layer.

3. To protect the bed during winter, spread 7–10cm (2¾–4in) of woodchip over the compost.

4. Leave the woodchip in place until you want to plant crops in the bed, as this will prevent weeds taking hold. Rake off the woodchip before planting – you could use it to make a path around your bed.

MAINTENANCE

Once the bed is set up, it will need an annual mulch of at least 5cm (2in) of compost every autumn or early spring to keep it productive. Remove any invading grass and remove weeds as soon as they appear. When compost is in short supply, give plants a liquid feed (see pp72–75) and spread fresh comfrey leaves over the surface for a quick nutrient boost.

SPACE TO SPARE?

If you can't use the whole growing space, cover any areas of bare ground with a couple of sheets of cardboard weighed down with stones. This will prevent weed growth and maintain soil quality until you are ready to plant.

START CROPS ON THE WINDOWSILL

Your windowsills could become your most useful space. They allow you to start growing early in the season, increase the amount of produce you can get from a small outdoor space, and grow tender crops during the winter months.

Use your windowsill from late winter to summer to raise plants from seed until they are ready to transfer outside. This means your growing season can start earlier, the plants will be stronger, and the period from seeding to harvesting is shorter, so that you can get more food from your outdoor plot.

A SAFE SPACE

One of the best aspects of a windowsill is the shelter that it offers. I use my windowsills to get frost-tender courgettes and tomatoes started well ahead of the last frosts outside. I get better harvests because plants can be transplanted as soon as the last frost is over. The fruit then develops earlier in the year, has more time to ripen, and cropping is extended, too.

Windowsills don't just provide protection from the elements. Many young seedlings are targeted by pests, such as mice or slugs and snails. Starting seedlings inside protects them while they grow through their period of vulnerability, and makes them less attractive to slugs once they are moved outside.

This protection isn't just for the seedlings, but for me too! It can be tough forcing myself to go outside to check seedlings on a rainy or stormy day, but if they are on my windowsill then I don't have to. I keep an adapted water bottle (see p49) on the windowsill next to them, so that if they ever look a little dry, I can water them there and then.

SUCCESSION PLANTING FOR SUCCESS

Starting seedlings on windowsills is also the key to succession planting. This is a system whereby you start growing one crop on the windowsill before there is room for it in your outside growing space. When an outside crop is harvested, the indoor seedlings can immediately be transplanted into the empty space. This effectively gives you an extra month of growth, as the initial development of one crop takes place inside, at the same time as another crop reaches maturity outside. This practice is hugely beneficial in cooler climates because it is a way of extending the growing season.

PLANT SWAP SPACE

Finally, windowsills offer a space for you to start a range of seedlings in containers to take to your local plant swap (see pp86–87). This is one of the most effective ways of growing excess seeds and swapping them for plants that you haven't got yourself.

Seedlings growing on windowsills are protected from the elements, as well as pests and diseases, in their earliest, most vulnerable stage of development.

TIP

If your seedlings start to bend towards the light when they are on the windowsill, rotate the containers 180 degrees and this will help straighten the growth.

TIP

Is your garden big enough to offer a small section to a friend who doesn't have their own space? Pooling resources and sharing tasks will result in even greater productivity.

ORGANIZE YOUR SPACE

Being able to grow food in your own garden is a huge privilege as well as a fantastic opportunity. Not only are you in control of the space, but you can organize it to suit your needs and grow the food you like to eat.

It is, however, a good idea to bear in mind other demands on that space before you get carried away and dedicate all of it to growing food crops. You may want an area of lawn where children can play, or a sunny, sheltered spot for sitting and reading. Both requirements can be successfully combined with growing food by, for example, having vegetables in raised beds around the perimeter of the lawn, or in containers around a seating area.

PRIORITIES AND TIMING

Use the SWAGA checklist (see p14) to establish your garden's potential, then consider how to split it into different sections for a setup that suits you.

- Grow salads close to the house for easy access, along with container-grown herbs to use in cooking.
- Establish your composting setup in a quiet, shaded corner with space along one of the boundaries to expand it over time.
- Collect rainwater from the roof via the downpipe and set aside a large bucket for storing greywater.
- Perennials are best sited at the far end of the garden because they need less maintenance.
- Ideally, grow annual vegetables along a south-facing boundary to make the most of the sun. If that isn't possible, an east-facing boundary is the next best option.

- Make sure you have space for an outdoor seating area for sitting and relaxing – perhaps opposite the annual veg section so you can observe your crops.
- Plants grown in pots or seedlings in modules ready for transplanting are best sited near to the water supply because pots dry out faster than raised beds or beds made in the ground.

By all means use the suggestions I've listed as a guideline, but remain flexible and adjust your setup if any sections aren't working, or if your requirements change. Before you set anything in stone, however, I recommend you prioritize two essential elements – water supply/storage and compost setup – and get these established so you can hit the ground running.

If your number one priority is to produce as much food as possible, don't be afraid to take the first growing season slowly. It is easy to underestimate the time and energy needed to grow food on a reasonable scale, so break the space down into sections and spread the work out over a realistic period of time. This will not only be less of a struggle than trying to get everything sorted at once, but it will also give you time to think about your long-term growing strategy (see p182).

I've covered my space with woodchip in order to suppress any weeds and the setup is quite varied. I prefer to grow some crops, such as courgettes, in containers, and others, such as salad leaves, in raised beds.

MAKE YOUR OWN TOOLS

Turning old items into useful gadgets for the garden is not only satisfying, but also costs virtually nothing. The heads of household brooms and brushes, for example, tend to wear out but the wooden handles can easily be repurposed.

The five DIY devices listed below are quick to make and easy to use. You will probably have most of the components to hand or will be able to source them for free.

a. Dibber

A dibber minimizes soil disruption and is very useful for making holes when sowing larger seeds (such as beans) and transplanting seedlings (such as leeks). The wooden handle left over when a spade head has broken off is perfect. Carefully use a sharp knife to shape the end to a blunt point and you have a long-handled dibber that saves on bending.

b. Seed trench maker

This simple, time-saving device is ideal for creating shallow trenches in raised beds or the ground and will quickly become indispensable. Take a piece of thick bamboo (or an old wooden broom handle) at least 2.5cm (1in) in diameter, measure the width of your bed, and cut the bamboo so it is slightly narrower. To use, press the bamboo down to the depth that you want the trench to be, then remove to leave a straight, instant seed trench.

c. Compost and water scoops

Large (2 litre/3½ pint) plastic milk containers make excellent scoops, because they already have a handle and a watertight cap. Repurposing these long-lasting and lightweight containers couldn't be easier. Simply cut off the base of the carton at an angle with a pair of scissors and your scoop is ready to use.

d. Measuring device

Annual and perennial crops need to be sown or planted at different distances, depending on the eventual size of the plant, and a homemade solid ruler will help you get these right. You will need a straight piece of wood around 60cm (2ft) long (such as a cut down pallet plank, stick, or section of broom handle), a ruler, and a permanent marker. Holding the ruler against the wood, simply mark out the different measurements to scale. Over time you'll find yourself using this device less frequently as your confidence about different crop spacing grows.

e. Seedling mister

The volume and speed of flow from a regular watering can are too much for fragile seedlings and could damage them. For a much finer spray of water, turn an old bottle with a spray nozzle into a mister to give your seedlings a very light shower. Empty spray bottles for non-toxic household cleaners make excellent misters but make sure you wash them out thoroughly beforehand.

a

c

d

b

e

MAKE YOUR OWN WATERING DEVICES

Seeds and seedlings are fragile, and need a light, gentle watering, whereas mature plants require more of a soak. You can easily make your own watering devices that reflect the needs of your different plants.

The three homemade devices below will cover the watering requirements for all of the plants mentioned in this book at each stage of their development. They are all lightweight, and each delivers water at a different rate, so which one you use will depend on the needs of the plant you are watering.

THE BUCKET WATERER

This device is the no-cost version of a watering can with a rose. You will need two buckets of equal size, or one bucket that fits into another, slightly larger one; the smaller bucket should have a handle. The buckets I use are from packs of suet balls for birds, and were sourced from friends and neighbours with bird feeders.

1. Gently drill holes in the base of the smaller bucket, or either of the two if they fit into each other. If you want to use the bucket to water salads and seedlings, drill lots of narrow holes, whereas if you want to water anything else, drill a few larger holes (make one of each type if you have enough buckets). Borrow a drill if you don't have one; don't use a hammer and nail because the bucket will crack.

2. Fit the bucket you drilled holes into inside the intact bucket, and fill it with water. I like to submerge both buckets in a larger container that has been filled with water, but you could simply pour in water from another container, such as a milk carton.

3. To use, carry both buckets around the growing area, taking out the second bucket and holding it over your plants to water them (*see opposite*). Place the filled bucket back inside the intact bucket when you have finished watering each part of you patch.

Punch holes in an old water bottle to create a seedling waterer.

MILK CARTON "WATERING CAN"

Empty milk cartons make perfect watering cans because they are free to source, come in different sizes, and already have a carrying handle. They are also lightweight and, unlike watering cans, have screw caps, which means you don't spill any of your precious water when carrying them around. Milk cartons can't, of course, hold as much water as watering cans so fill up three or four at a time when using them. The "carton can" is ideal for watering individual perennial and annual plants and large seedlings. It isn't suitable for watering young seedlings because the strong flow of water may cause damage.

SEEDLING WATERING BOTTLE

A repurposed plastic bottle that delivers a fine spray to fragile seedlings is one of the most useful pieces of equipment to have at hand and will last for years. To make, take an empty, washed 500ml (16fl oz) bottle and punch five to seven very small holes in the cap using a nail – tapping gently with a hammer – or the sharp point of a compass (one you'd find in a geometry set). Fill the bottle with water, screw on the cap, and it's ready to use. You can also use the seedling mister on p46.

3

PRODUCE YOUR OWN COMPOST

HOW TO MAKE NUTRIENT-RICH
COMPOST FOR FREE

SOIL AND COMPOST: THE ESSENTIALS

These two ingredients are vital if you plan to grow your own food. Although similar in appearance, each has its own benefits and requires a significantly different length of time to produce.

There are several different types of soil, but I like to keep things simple. As a general rule, when I talk about "soil", I am referring to "topsoil", the brown material you find just under a lawn, for example. Topsoil is made up of organic matter (humus) microbial life, mineral particles, air, and water, and has a high percentage of nutrients.

Below the topsoil is the subsoil, which is much lighter in colour. This has less organic matter and microbial life than topsoil, and sometimes has large stones mixed into it.

Soil forms after centuries of weathering has broken down stone particles. It can take about 200 years to produce 1cm (½in) of topsoil and it is considered to be a non-renewable resource.

WHAT IS COMPOST?

Compost is made from decomposed organic material, is rich in nutrients, and is a darker brown than soil. Compost also encourages beneficial microbial life to thrive, which research has shown to be the key to healthy, disease-resistant plants, especially if you follow the low- or no-dig methods I use in this book (see p68).

Good garden compost can be made in as little as three months with only grass clippings, shredded paper, and shredded autumn leaves, but I prefer to work with nature and produce compost in yearly cycles using a wide range of ingredients (see pp64–67). This ensures that my compost is well broken down and contains the right balance of nutrients. A year might seem a long time, but it's nothing compared to the time required to produce soil.

COMPOST VERSUS SOIL

It's possible to grow plants just in compost or, in the short term, just in soil (see pp38–39). If I had to choose one of the two materials to grow crops in, I would opt for compost. This is because it tends to contain more nutrients for plants to absorb than soil.

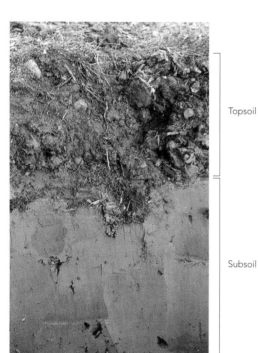

Topsoil

Subsoil

The different properties of topsoil and subsoil are clearly visible in this cross-section.

Soil (*left*) **and compost**
(*right*) look very different.
Compost is darker with
visible lumps, whereas soil
is lighter and finer.

GET THE RIGHT BALANCE

The best growing medium is a mixture of compost
and soil, ideally a 50/50 mix, as plants grown in soil
enriched with compost will almost always be more
productive than those grown in soil alone. However,
at the beginning of your journey, you may not have
had time to make compost, so you can use extra soil
as a bulking agent. This will help to ensure your
precious compost lasts as long as possible until you
have established a large enough supply that you no
longer have to ration.

TIP

*If you want to grow crops
in compost alone, make sure
that the compost you use has
been made from a wide range
of ingredients. This will
ensure that it has a good
balance of nutrients.*

USING HOMEMADE COMPOST

There are so many uses for compost around the garden,
from helping young plants get a good start, to ensuring mature,
established plants continue to produce strong yields.

Think of compost as fuel for your garden: a car won't work without fuel and neither will a garden work without compost. Be aware that some homemade compost might contain weed seeds, so be vigilant and remove any weed seedlings that pop up. If you can't easily tell the difference between a weed and a vegetable seedling, wait until the first pair of true leaves appear, so it's easier to work out which are weeds. If you have a neat row and a seedling is growing outside of it, it's likely that the odd one out is a weed.

SOWING SEEDS
Homemade compost is the perfect growing medium to sow seeds into. Many vegetables are annuals and need be sown from scratch every year. That's a lot of seed-sowing! Special compost mixes with various different ingredients are often recommended for sowing seeds, but all you really need is compost. I've been sowing seeds in homemade compost for 15 years and it works. If, for instance, you are sowing dried peas from the kitchen, just fill cardboard rolls with your own compost and insert the peas.

To sow fine seeds, such as lettuce or tomatoes, some people suggest sieving compost before using it. Personally, I never bother to do this because the seeds will still grow for you. You can sow directly into compost in raised beds, containers, or small pots.

POTTING ON AND TRANSPLANTING
Your seedlings will soon need potting on – another use for your homemade compost. For example, brassica or lettuce seedlings sown close together should be separated out, so taht they have space to grow. I plant the separated seedlings into individual pots filled with my compost. Other seedlings may need a bigger pot (and more compost) to keep them going until the weather warms up. If, for example, you have sown tomatoes and spring starts later than expected, you will need to pot the seedlings on once or twice to keep them growing.

Transplanting seedlings into the ground or raised beds outside can also require a little compost. Unless you are transplanting directly into fresh compost, it's always good practice to add a bit of compost to the hole when planting out a seedling. This gives a boost of nutrients and welcomes the plant to its new home.

When I transplant annuals, such as kale, into a raised bed, I always put a handful of compost at the base of every hole. Get into the habit of adding compost when transplanting and your plants will thank you. Perennials, especially, will need the best possible start, so when planting a young blackcurrant bush, for example, make the hole twice as big as the rootball and always add compost at the base and sides of the hole.

IMPROVING AND PROTECTING SOIL
Once a year, I top up my raised beds and containers with 5–7cm (2–3in) of compost to replenish nutrients used up by the previous crops (see p156). Mulching raised beds or perennials annually encourages a "soil web" of beneficial fungi and microbes to help facilitate strong plant growth and disease resistance. This organic matter is also fantastic at retaining moisture and can keep plants healthy during prolonged periods of low rainfall.

Compost has many uses, for sowing seeds (*top left, above left*), transplanting (*top right*), and improving soil (*above right*).

GETTING BY WITHOUT COMPOST

Growing crops without compost in your first growing season isn't as hard as you might think. Use these four substitute mulches to add essential nutrients to your soil.

- A thin layer of grass clippings once every few weeks
- 1cm (½in) layer of coffee grounds every month
- Water plants with diluted liquid feed every month (see pp72–75)
- Add a dusting of wood ash over the soil each year

YOUR COMPOST BIN

According to a popular gardening saying, the best time to plant an apple tree was seven years ago; the second-best time is now. This principle can also be applied to starting a compost bin, but at least it only takes a year to catch up.

If you are aiming to produce an abundance of food, especially when you are growing food for free, a composting setup is vital. You may decide you need one, four, or even more compost bins, but creating your own compost is the single most important thing you can do.

FILLING IT UP

Start now and focus your energy on filling your first compost bin as soon as possible. The sooner you get going, the sooner you can leave the contents to decompose, and in a year's time (or less), you will have your own beautiful homemade compost. Use resources from your own household and source free supplies from elsewhere (see pp64–67).

Once you have created and filled your first bin, you will have become more knowledgeable about the best places to source composting materials in your local area, as well as the best times to collect it. This will speed up the process of filling additional bins.

HOW LONG WILL IT TAKE?

A year may seem like a long time, but don't worry too much about not having any homemade compost in the first year. It is possible to grow plants without compost, as long as you have some topsoil (see p38), especially if you focus on perennials in the first year (see p78). I recommend that in the first year, once you have filled your compost bin, you turn its contents every month. This way you should have usable compost within six months, and hopefully in time to do the autumn mulches for which it is essential.

Alternatively you can take a different approach and simply concentrate on getting all your infrastructure in place before you start sowing and planting. Then you can begin to grow in earnest during your second year, when your compost will be ready to use.

The good news is that wherever you site your bin, you *will* produce compost – it's just a matter of time. The best place to put a compost bin is in a sheltered sunny location as the heat will help speed up the decomposition. However, sunny sheltered spots should be prioritized for growing crops. This means it's often best to place your bin in a shady part of the garden, to avoid taking up vital food-growing space. If you have a large garden, this probably won't be an issue for you.

Quantity is also important. The bigger the bin and the more you put in it, the faster the composting process happens. This is because heat speeds up decomposition, and a large pile of compost material heats up and retains heat more easily than a small one.

Compost is the currency of the garden and will help to ensure your plants grow and thrive. You can never have too much compost – the more you have, the more you can grow – just be sure to ration your compost if you are running out, prioritizing it for when you sow seeds and mulch. You can also bulk out your compost with leaf mould (see p59) or topsoil.

Turn the contents of your compost bin every month or two of months to speed up the breakdown of materials.

BUDGET COMPOST-BIN SOLUTIONS

It's possible to make compost in many different types of compost bin.
Even if you only have a terrace or balcony, there are smaller solutions
that you can use. Pick the one that suits your space best.

There are a variety of styles of compost bin that you can easily make using found or low-cost materials. Here are some to try building yourself.

a. Pallet-plank bin
I produce my compost bins using planks from broken-up pallets and reclaimed wooden posts. They are very simple to put together (see pp60–61).

b. Pallet-only bin
To make this simplified version of my pallet-plank bin, just bind four pallets together with rope or wire. If appearance isn't an issue and you need an instant option, this is the bin for you.

c. Wire bin
Old chicken or fencing wire can be repurposed to make a very simple compost bin. Use four posts as a frame and wrap the wire around them to create a square bin, or just go for a circular style.

d. Woven bin
This is a great, low-cost solution if you live near a good supply of branches. Drive eight or nine thick, 1.2m (4ft) branches vertically into the ground in a 1m (3ft) diameter circle, and weave thinner branches between them horizontally to create the sides. Hazel and willow work really well for this.

SHORT OF SPACE?
If you only have a small space, particularly if you live in an apartment with a balcony, the low-cost solutions suggested above may be too large. Compost takes longer to break down in small bins, but you will still be able to make a good amount.

For use in smaller spaces, you can convert a small dustbin or 25-litre (5½-gallon) bucket with a lid into a compost bin by drilling holes in the base and sides for ventilation. Sit it on a few layers of cardboard or raise it on bricks if you only have a hard surface to put it on. If your compost starts to smell bad, mix in lots of brown materials and check back the next day. If there is still a foul smell, add more. This should solve the issue.

LEAF MOULD
A wire bin is great for making leaf mould, which has excellent moisture retention and can be mixed with compost for sowing, added as an extra material in a new raised bed, or used as a mulch for perennials. Large quantities of dry leaves are often readily available in autumn. Just fill your bin and allow the leaves to rot for at least two years before you use them. Mixing in some coffee grounds will help speed up the process.

BUILD A COMPOST BIN FOR FREE

You can make a strong, functional bin using a few pallet planks and four pieces of reclaimed wood, such as old fence posts. Wherever you live (rural or urban), these materials are easy to source and often given away for free.

HOW TO BUILD

There are variety of places to source pallets for free (see p36). You'll need two to three pallets to make a 1m (3ft) tall bin like this one. Try to source them well ahead of time, so that you can leave them out in heavy rain for a few hours. This will soak the wood, making the pallets easier to dismantle. I used screws I had lying around to build this bin. If you can't find any, you could use the reclaimed nails from breaking down your pallet instead (or even a mixture of screws and nails).

1. Split down your pallets until you have the required materials (*see opposite*). Wear eye protection and gloves, and use a bit of brute force. As you split the pallet

into individual sections, remove any nails using the claw hammer and keep them in a pot. Before you start, lay out all of the wood to ensure you have everything.

2. Measure the length of the pallet planks and put two posts flat on the ground the same length apart. Lay a plank across the two posts at one end. The outer edge should be flush with the plank's edges, with no over-hang. Screw the first pallet plank to the posts. Repeat at the other end of the posts with the second plank, then with three more planks to create a slatted side.

3. Lay the other two posts on the ground as before and repeat step 2, using five more pallet planks and 10 screws to create the second slatted side. Set down

your two complete side panels, posts facing down as shown. Place a set of five pallet planks beside each, laid out at the same intervals. You should have two sides and 10 planks.

4. Ideally, find a second person to help you join the two panels. It's a little tricky to do on your own, but it is possible (with a bit of care, a cordless electric drill, and a piece of flat ground). Upend one of the side panels, with the post lying on the ground and facing inwards. Use a plank to keep it upright if it's unsteady. Repeat with the other side, so the distance between the two is the length of the pallet plank. Screw the first plank down so it joins the top of the two sides (with no overhang) then screw a second plank down to join the base. Repeat with the remaining three planks.

5. Roll the three-sided pallet structure over so the fourth, open side is facing upwards. Screw the five remaining pallet planks down, as in step 4. This should be a lot easier because the structure is more stable. You now have a complete compost bin.

TOOLS AND MATERIALS

You will need gloves, eye protectors, a pry bar, claw hammer, and screwdriver (cordless electric if you have one; if not, try to borrow one) plus:

- 4 reclaimed posts around 1m (3ft) tall, such as old wooden posts or door frames
- 20 pallet planks, all the same length
- 40 screws

4   5

COMPOST RECIPES

Making high-quality compost isn't a case of throwing all of your waste onto a pile and leaving it; you need to add the right ratio of different materials.

Compost ingredients can be split into two basic categories, brown and green materials, both of which are crucial for a healthy, problem-free compost bin. Brown materials are carbon-rich substances, which are usually dry and so are often lightweight – these provide the building blocks for plant matter. Green materials are all the nitrogen-rich materials, which are usually heavier than brown materials as they also contain moisture. Nitrogen is required for healthy plant cell growth and function.

Don't be fooled by the actual colour of compost ingredients: used coffee grounds may look brown, but they are classed as green material because they contain a lot of nitrogen. See my list of different materials (*opposite*), and in more detail on pp64–67.

TWO PARTS BROWN TO ONE PART GREEN

This couldn't be an easier ratio to remember; just add two full buckets of brown material to your bin for every bucket of green. Stick with this ratio and you are unlikely to have problems.

I tend to build my compost bin up in layers, like making a lasagne (*see opposite*). But it really doesn't matter if you mix different greens or browns together in the same layer. You will have various types of material available at different times of year, and diversity is a definite advantage.

Generally speaking, as long as you stick to the right ratio, the more diverse your compost materials, the better the quality of your compost. You can keep recipes simple – just ensure you use the right proportions of each colour.

MAINTAINING YOUR COMPOST BIN

A couple of months after filling a compost bin, I like to turn the contents to combine all of the ingredients and incorporate air. I use a fork to lift up the material at the bottom of the bin and place it on the top. It's not strictly necessary, but it does speed up the process. If you use fine materials (such as shredded newspaper) and turn them each month, you could be looking at garden-ready compost in six to eight months. However, most compost will take around a year to achieve a good texture and quality.

TIP
Once your first bin is full, and if you have space, why not start on a second?

WHAT TO PUT IN YOUR COMPOST BIN

It's important to know which materials are "green", which are "brown", and which to avoid composting.

Green materials

- Fruit and vegetable scraps
- Softwood plant prunings
- Weeds without seedheads
- Rabbit, chicken, duck, and guinea pig droppings
- Horse and cow manure
- Grass clippings
- Used coffee grounds

Brown materials

- Newspaper, paper (not magazines), and cardboard
- Sawdust (from untreated wood)
- Autumn leaves
- Dried grass
- Dead plant matter
- Deciduous wood ash

What not to compost

Some things are just not suitable for domestic compost because they can, for example, attract rats, leach chemicals, or emit unpleasant odours. Here are materials to keep out of your compost bin:

- Cheese, milk, and other dairy products
- Meat and bones
- Dog and cat faeces and litter
- Weeds that have gone to seed
- Diseased plant materials
- Glossy paper

Green layer

Brown layer

Autumn leaves

Coffee grounds

Newspaper

Fruit and veg scraps

Dried grass

Grass cuttings

The lasagne method works by adding layers of green and brown materials to your compost bin one after the other.

SOURCE FREE COMPOST MATERIALS

You can never have too much compost! These are materials you can use to fill your compost bin – look for them at home first, and further afield if you need to. I've suggested some locations and seasons when they are most readily available.

Remember, you don't need all of these items to make good compost – whatever you have available will work, as long as you stick to the right proportions (see p62).

a. Fruit and vegetable scraps
Uncooked fruit and vegetable scraps are high in nutrients, so compost any of this waste you have at home. Collect scraps from neighbours, or ask a local restaurant or pub. If they are willing to help, give them a container to fill. Avoid unsuitable items (see p63).

- **What** Green material for nitrogen
- **Where** Urban, suburban, and rural
- **When** All year round
- **How** Add in layers up to 10cm (4in) thick, or mix with a brown material, such as shredded paper.

b. Horse and cow manure
If you keep animals, you will have plenty of access to this, but otherwise try bartering your home-grown produce for manure. Many farms and stables will have large piles of well-rotted manure for you to take from, and you'll be able to source a large quantity if you make the right connections. Try to find a place where animals are fed with organic crops or natural feed, as pesticides can persist in manure when animals have eaten feed that has been sprayed.

- **What** Green material for nitrogen
- **Where** Rural
- **When** All year round
- **How** Add manure (no straw) to your compost bin in layers no thicker than 5cm (2in).

a   b

LEAVES SHREDDED WITH A LAWN MOWER WILL BREAK DOWN A LOT MORE QUICKLY THAN THOSE COMPOSTED WHOLE.

d. Lawn clippings

Grass clippings may be one of the easiest compost ingredients to source – especially if you have a lawn that you regularly mow yourself. Ask neighbours with gardens for their clippings if you don't have a lawn.

- **What** Green materials for nitrogen
- **Where** Suburban and rural
- **When** Early spring to late autumn
- **How** Add layers no more than 3cm (1¼in) thick so they don't form a sludge.

c. Autumn leaves

These are a fantastic free resource, but one that can disappear if you don't collect them quickly. Start with those in your garden, and if you still need more, try collecting them from the sides of quiet country roads and pathways. You can also look out for bags left outside by suburban garden owners who have no use for them. In urban areas, speak to the local council about collecting some leaves from local parks.

- **What** Brown material for carbon
- **Where** Urban, suburban, and rural
- **When** Mid- to late autumn
- **How** Leaves compost fairly slowly, so you can either make a wire bin (see p59) or add to your compost pile in thin layers.

TIP

Use only grass clippings from a lawn that hasn't been sprayed. Lawn chemicals will kill beneficial microbes in soil and could negatively impact plant growth.

c

d

Shredded paper

If you or a family member work from home, collect any shredded paper. Alternatively, approach local solicitors, accountants, and other office environments – who often have a huge quantity of paper to dispose of. Just remember to avoid glossy paper, which may contain chemical pigments.

- **What** Brown material for carbon
- **Where** Suburban and urban
- **When** All year round
- **How** Add in layers of up to 5cm (2in)

e. Newspaper and cardboard

If you don't buy newspapers or receive many parcels yourself, these materials are easily sourced from pubs, hotels, restaurants, and independent corner shops. Most newspapers use water- and soy-based inks. Inks that don't smudge when rubbed are safe to compost; a dark, oily smudge indicates petroleum-based inks, which must be avoided.

- **What** Brown material for carbon
- **Where** Urban, suburban, and rural
- **When** All year round
- **How** Tear into small pieces and add in layers up to 5cm (2in) thick

Human hair

You may be surprised to find out that it's fine to add human hair to the compost bin. It is high in nitrogen, so you don't need much. Approach local hairdressers for supplies.

- **What** Green material for nitrogen
- **Where** Urban, suburban, and rural
- **When** All year round
- **How** Sprinkle in thin layers

f. Rabbit, chicken, duck, and guinea pig litter

These can safely go in the compost bin – the manure contains plenty of nutrients. Find out if you have animal- or pet-owning neighbours who would otherwise be throwing out used bedding. Ideally, the animals' food will be 100 per cent natural and the bedding chemical-free.

- **What** Brown and green material for carbon and nitrogen
- **Where** Urban, suburban, and rural
- **When** All year round
- **How** Add to compost in layers of up to 20cm (8in)

e  f

g. Sawdust and woodchip

Sawmills are a fantastic source of sawdust and wood shavings to add to your compost pile. For woodchip, search out local tree surgeons who are usually happy to offload shredded material. You might also spot small piles of free woodchip on the side of roads where tree work has been carried out.

- **What** Brown material for carbon
- **Where** Suburban and rural
- **When** All year round
- **How** Sprinkle sawdust into the compost when you apply other materials. Be sparing with woodchip and use only small chips in thin layers.

h. Used coffee grounds

Coffee grounds as a compost ingredient have received some bad press, but I've been adding them for years with good results. Grounds are high in nitrogen, the pH is almost neutral, and the material is fine and breaks down quickly. Collect them from your own machine, and take an empty tub to fill at your local café.

- **What** Green material for nitrogen
- **Where** Urban, suburban, and rural
- **When** All year round
- **How** Sprinkle on compost or add a thin layer.

APPROACH YOUR NEIGHBOURS

For a regular supply of household waste ingredients, such as vegetable scraps, cardboard, leaves, and grass clippings, neighbours are your best source. If you have recently moved to the area, introduce yourself, and make your request via a note through the letterbox. For food scraps, supply each neighbour with a clean container, ask them to leave it outside, and organize collection. Offer to collect cardboard, leaves, and grass clippings in person at a convenient time. In return, an offer of produce (or even plant cuttings) will always be welcome.

TIP

If you want your leaf mould to break down faster, mix in several handfuls of used coffee grounds.

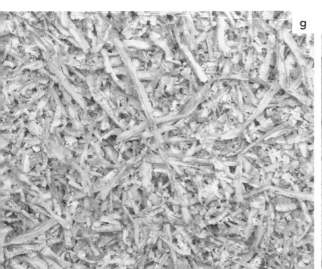

g

h

PERMACULTURE

A system inspired by nature, permaculture is at the heart of how I grow food.
By following its ecological principles, you can be confident that whatever you
grow will be produced using the most natural and sustainable methods possible.

The word "permaculture" is an amalgamation of "permanent" and "agriculture", and was coined in the 1970s by two Australians, Bill Mollison and David Holmgren. Their aim was to create sustainable, food-producing landscapes using nature as inspiration, and Holmgren subsequently came up with the 12 principles of permaculture. These are thinking tools to help create sustainable designs, but don't let the word "design" put you off. It can be applied to something as simple as finding the best location for a new raised bed.

PUTTING PERMACULTURE INTO PRACTICE

My favourite permaculture principle is "observe and interact", which I use time and time again. I take a step back, look closely at what is happening in the garden over a certain period, and come up with ideas for improving the space and utilizing any opportunities to the full. On a simple level, this could mean observing the movement of the sun over a day, establishing where the shadiest corner is, and creating a compost setup there.

In practice, permaculture is about working with nature rather than against it. I have been growing food organically for more than 15 years and neither wanted nor needed to use chemicals, such as fertilizers or insecticides. Instead, I have put strategies in place that keep soil and plants healthy, which I discuss in this book.

NO-DIG GARDENING

The practice of no-dig gardening to grow food, popularized by Charles Dowding, follows the principles of permaculture because it is a method inspired by nature. Think of the natural cycles at work in a deciduous forest. Every autumn, the leaves fall to the ground where they decompose and, with the help of microorganisms, return nutrients to the soil. These nutrients are then taken up by trees to promote growth and maintain health. Forests are self-sustaining and recycle their resources, which enables them to thrive for centuries without additional sources of nutrients.

A garden does, however, differ from a forest in one important aspect: we are removing nutrients from the soil in the form of harvests. We must, therefore, add compost (organic matter) to the soil to top up nutrient levels and keep it healthy. Traditionally, gardeners used to dig organic matter back into the soil, but this not only destroys its structure, it also disturbs the worms and microorganisms that live there. Spreading organic matter over the surface, where it will decompose naturally and feed the soil, is much more effective.

To summarize, here are three key reasons to use the no-dig method:

- **Moisture retention**
  A growing medium with its natural structure intact captures and retains more water than soil that has been dug over. You will reap the benefits in hot weather and drought.

- **Weed reduction**
  Digging organic matter into the soil simply brings weed seeds up to the surface to germinate. Adding a layer of organic matter as a mulch on the surface means that far fewer weeds will germinate.

- **Free and simple**
  Once you are producing your own homemade compost, adding an annual layer to your growing area to feed the soil costs nothing. Spreading it over your growing area is much less effort than digging it in and does no damage.

Homemade compost should smell earthy, just like the smell of a forest floor. If it doesn't smell good, then it isn't ready to be used directly (but can be used for mulching and filling beds in autumn).

WOODCHIP AND BIOCHAR

Of these two materials, woodchip makes an excellent mulch and will improve the moisture retentiveness of your soil. Biochar is charcoal that, once activated, adds nutrients and structure to the soil. Even better, both can be sourced for free.

WOODCHIP

A fantastic free resource for gardeners is woodchip, as it can be used as a mulch. When woodchip breaks down, it releases nutrients that contribute to a healthy soil structure.

The best woodchip for mulching and composting is created from chipping large, freshly pruned branches; this is sometimes called "dirty woodchip". Many gardeners will wait for pruned branches to dry out and the leaves to drop off before feeding them into a woodchipper – I call this "clean woodchip".

The easiest way of sourcing dirty woodchip is to ask a local tree surgeon for some of theirs. They may have to pay for their woodchip to be taken away, so are often more than willing to part with it for free (I know many people who have been collecting woodchip from local tree surgeons for free for years). If you live in a rural area, a neighbour may have their own woodchipper. If they do, ask if you can take their chippings, or even borrow the woodchipper itself.

Only use woodchip from native broadleaf trees, as woodchip from softwood, such as pine, can be very acidic and can negatively affect your crops' growth.

PREPARING AND USING WOODCHIP

Compost dirty woodchip before using it to grow annuals. Place any you have in a large compost bin and leave it for two to three years, turning once each winter and summer. By this point, you will have a beautiful moisture-retentive mulch to use around your annuals. You can also use your composted woodchip in a 50/50 mix with compost to sow seeds and transplant seedlings. Alternatively, you can use dirty woodchip right away as a mulch for your perennials.

Clean woodchip is perfect to use direct as a mulch for perennial fruits in autumn and spring – it will keep

Dirty woodchip is a partly decomposed mix of wood, bark, and leaves. It is created when tree surgeons feed fresh branches into a woodchipper.

weeds down and maintain moisture levels in your growing medium. Use it for perennials in pots, raised beds, or directly in the ground. Once every year or two, add a 5cm (2in) layer of compost or well-rotted manure and a couple of handfuls of wood ash around the base of the perennial fruit before mulching with 5–7cm (2–2¾in) of woodchip to give a boost of nutrients. You can also use clean woodchip to make pathways around your growing space.

5CM (2IN) OF WOODCHIP WILL SUPPRESS MOST WEEDS AND RETAIN SOIL MOISTURE.

BIOCHAR

There are conflicting views about the suitability of these small pieces of carbon-rich charcoal as a soil additive. From reading countless studies and seeing citizen trial results, I believe that activated biochar (biochar soaked in a nutrient-rich liquid) provides a good means of improving the fertility of poor soils. This is because having absorbed nutrients from the liquid feed, biochar slowly releases them into the soil. Biochar also provides an excellent habitat for beneficial soil microorganisms, as it is porous and retains moisture.

SOURCING CHARCOAL

The best place to find charcoal is where people have been burning wood on a fire. This could be a campsite or fireplace – the bigger the fire, the better. Avoid fires where plastics or non-woody biomass have been burnt, and don't use barbecue charcoal, as this often contains unwanted contaminates. Charcoal is created by pyrolysis – high-temperature combustion without the presence of oxygen. It is a dark black colour, feels very brittle, and can be easily crushed.

Collect the pieces, put them in a strong bag, and use a stone to crush them into smaller fragments. Your biochar is now ready to be activated.

ACTIVATING AND APPLYING BIOCHAR

All you need is a bucket, water, and something to use as an activator, such as diluted liquid plant feed, a few handfuls of compost in water, or even undiluted urine. Place the biochar in a bucket, cover it with water and activator, and leave for around a week.

To use activated biochar, mix it into your growing medium at no more than one part biochar for every nine parts growing medium. You can use this mixture to fill pots or raised beds.

Alternatively, if you have a good supply, use biochar as a composting ingredient without needing to activate it. Just add a handful to your compost bin every now and again, and the biochar will naturally be activated by the compost over time.

Activate biochar with your chosen activator (*left*) and mix it into your compost or lightly mulch your beds and containers with it (*above*).

MAKE YOUR OWN LIQUID AND COMFREY FEED

Liquid feed can be a quick fix if plants are looking a little undernourished, or a regular necessity for container-grown fruiting plants in flower. Directing nutrients in a liquid form straight at the roots of the plant makes them far easier to absorb.

I make my own liquid and comfrey feed (see *opposite*, and pp74–75) because it is a useful substitute for compost and the plant material can be added to the compost bin later, so nothing is wasted. Liquid feeds give an instant boost to container-grown fruiting plants, such as tomatoes, or plants that look undernourished and have pale leaves; the goodness from a top-dressing of compost, on the other hand, will take some time to penetrate down to the roots. You can also use liquid feed on poor soils to give the plants growing in them an extra boost of nutrients.

Once you have a plentiful supply of compost, you'll be able to use this to mulch your containers and beds, and you'll have less need for liquid feed. However, you may still want to use it for fruiting vegetables and plants growing in containers.

APPLYING AND STORING THE FEED

Always dilute concentrated feed, using no more feed than one part to every seven parts water. Simply fill a watering can with diluted feed, and water the soil around the plants.

Store liquid feed in a cool, dark place to reduce algae growth in the bottle. Liquid feed doesn't go off (it's smelly enough to start with!), but I aim to use mine within a year.

PLANTS ABSORB NUTRIENTS IN LIQUID FORM FAR MORE QUICKLY AND EASILY THAN IN A SOLID FORM, SUCH AS COMPOST.

NUTRIENTS: N, P, K

Different materials used in liquid feed contain different levels of nitrogen, phosphorus, and potassium (NPK), and it's best to match the type of feed to the plant or to its stage of growth. So what exactly do nitrogen, phosphorus, and potassium do for your plants?

- **Nitrogen (N)** is the key nutrient for healthy, green leaf growth. Get it from nettles, grass clippings, and chicken manure.
- **Phosphorus (P)** is vital for all-round healthy plant development. Chicken and other animal manures are good sources.
- **Potassium (K)** aids fruit formation and growth. Sheep manure and comfrey are rich sources.

HOW TO MAKE LIQUID FEED

Making this liquid feed couldn't be easier but, be warned, it can give off a powerful odour! I've chosen to make nettle feed here, as it is particularly beneficial for leafy crops such as kale and spinach. However, you can substitute any of the ingredients given in the nutrients box (*opposite*) for nettles, depending on the crops or the reason you want to make feed.

1. Collect a bucket of nettle leaves (wear gloves and long-sleeves). Tear the leaves up and place them in the centre of the cloth spread out on the ground.

2. Gather up the sides of the cloth and tie them to the stick with the string, using a double knot. Rest the stick across the top of the bucket and fill the bucket with rainwater so that the bundle is submerged. Then leave the bucket outside, but sheltered from rain.

TIP

Seaweed contains many trace elements that plants need for healthy development. If you can responsibly source seaweed, add a couple of chopped-up handfuls to any liquid feed you make.

3. After a week, lift out the cloth bundle, give the liquid a stir, and then re-submerge the bundle in the solution.

4. At the end of the second week, remove the bundle and empty its contents into your compost bin or use it as a mulch around tomatoes or peppers. Decant the liquid feed from the bucket into bottles and store in a dark place.

TOOLS AND MATERIALS

You will need:
- nettles
- a bucket
- gloves
- an old tea towel or piece of cloth
- a stick that is longer than the diameter of the bucket
- string
- rainwater
- bottles

COMFREY FEED

Comfrey's impressive taproot (up to 3m/10ft long) mines nutrients deep below the surface of the soil and brings them up to its mass of leaves, resulting in foliage full of essential nutrients. Turning these lush leaves into a liquid feed is a great way to boost plant growth, especially for container-grown crops.

Diluted comfrey concentrate also makes a fantastic homemade tomato feed that is best applied while the plant is flowering and fruiting. If you have feed to spare, you could always keep it in glass jars to swap with other gardeners for new plants or seeds.

Comfrey is best grown in the ground. Keep it separated from the rest of your crops so that it doesn't take over your beds.

GROWING COMFREY

Comfrey is a perennial and is easy to grow from root cuttings. It is commonly found at plant swaps and local allotments. In early spring, plant a 5cm (2in) piece of root in a compost-filled pot, then transplant it into the ground in early summer.

Comfrey grows well in shade, but partial shade is best. Grow a variety called Bocking 14, as it is infertile and doesn't set seed. Once planted, comfrey can be hard to eradicate, so think carefully about where you plant it. On the flip side, you will have plenty of cuttings for new plants to give to friends.

DILUTING AND APPLYING CONCENTRATED FEED

Make up a mix of one part concentrated comfrey feed to 20 parts water, and water it straight onto crops such as beans, peppers, and cucumbers when their flowers form.

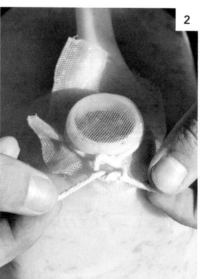

HOW TO MAKE COMFREY FEED

It will take around three to four weeks to produce a single jar of comfrey feed. When comfrey leaves decompose, they can smell even more unpleasant than nettles, but you can minimize any odours by following this method. Don't expect huge amounts of feed, but a little of this highly nutritious liquid will go a long way. The hairs on the leaves can be irritating, so be sure to wear gloves when you're handling them.

1. Turn the container upside down and cut off the base using sharp scissors or a knife (*see below left*).

2. Cover the opening at the top of the bottle with some sort of mesh fabric – such as hessian or an old sheet. You can secure it in place using some string.

3. Pack the container to the brim with with torn-up comfrey leaves.

4. You need to keep the container upside down, such as balanced on other objects or tied to a fence or post. Leave enough room underneath the neck for a glass jar to catch the liquid.

5. Once the jar is in place, put a brick or stone on top of the filled container to squeeze out the juice from the leaves.

6. Keep topping up with more comfrey leaves as they rot down and the level drops. After four weeks, squeeze any remaining liquid from the leaves while wearing gloves. When the jar is filled, seal with the lid and store in a cool, dark spot. I aim to use mine within a year.

TOOLS AND MATERIALS

You will need:
- a container with a neck (large bottle, milk carton, or 5-litre/1-gallon jug)
- scissors or a sharp knife
- a small amount of mesh fabric
- string
- gloves
- comfrey leaves to fill the container
- an empty glass jar with a lid
- a heavy brick or stone

SOURCE SEEDS AND PLANTS FOR FREE

HOW TO START OFF YOUR
FIRST PLANTS WITHOUT
SPENDING ANY MONEY

ANNUALS AND PERENNIALS

Annual crops produce food within a few months, while perennials take a year or more to mature and produce their first crop. A good plan is to prioritize sourcing perennials, so you can get them to maturity sooner.

All fruit and vegetable plants can be divided into two categories – annuals and perennials. Put simply, annuals are plants that mature and yield a harvest in the same year that they were sown, but need to be resown each year. Perennials continue to grow and be productive over many years. Growing both gives you a range of crops to harvest throughout the year.

In this book, growing instructions for annuals and perennials have been split into separate chapters. Before you start growing them, you need to be able to source seeds, seedlings, and cuttings for free.

STRATEGY FOR YEAR ONE

Perennials need a bit of time to establish well, so it's a good idea to get them planted as soon as possible. Rhubarb, for example takes three to four years to reach full productivity. For the first couple of years, the size of your harvest may not meet your expectations, but once the plants reach a good size, you can expect gluts for years to come. When fully mature, perennials are easy to propagate, so you'll quickly be able to increase your stock and the size of your harvests.

For this reason, I strongly believe that once you've decided to set up a space to grow food for free, your focus for the first year should be on perennials, with annual crops given a lower priority. Think of this as a long-term strategy, designed to achieve maximum yield in as little time as possible. Of course you should also grow annual vegetables, such as salads and peas or beans in the first year, but it's best to allocate space for perennials in your setup first, as they will be staying put. Once your perennials are in position, your annual crops can be planned around them. See pp88–115 for instructions on how to grow specific perennial crops.

ANNUALS OR PERENNIALS?

Annuals:

- Are more productive per square metre than the majority of perennials
- Have a short growing season, so another crop can replace them in a bed or container once they have been harvested
- Are a source of fresh produce in winter and before other crops are ready
- Include crops such as chickpeas and runner beans, which are a source of protein

Perennials:

- Are usually far simpler to propagate than annuals
- Develop strong root systems and are more resilient to extremes of weather
- Almost always produce reliable harvests, even gluts
- Are low maintenance and don't need to be replanted or moved each year

Strawberries are perennials, and once established, it's easy to increase your stock by potting up the runners at the end of the creeping shoots.

Pepper

Sprouted potatoes

Tomatoes

Potatoes

Garlic

Runner beans

Chickpeas

Peas

PLANTING FROM YOUR KITCHEN CUPBOARDS

The first place to look when sourcing free seeds or tubers to grow is your kitchen cupboards. Out-of-date food, such as sprouted potatoes and old dried peas, can easily be planted and will yield fresh crops.

Much of the food we eat is (or contains) a means of growing new plants. For example, potatoes are tubers, garlic cloves are bulbs, and peas are seeds – these are all items that can be sown to grow new plants. Fruit and fruiting veg (such as tomatoes and peppers) also contain quantities of seeds.

Below you'll find brief instructions on how to grow a selection of vegetables that are commonly found in kitchen cupboards and are easy to propagate. This will give you a range of new plants without spending a penny. More comprehensive growing instructions for these plants can be found in the general chapter on how to grow annual plants (see pp116–157).

Not every seed you find in your cupboard will be viable, so do a germination test on a sample of each seed type before you commit to planting an entire batch (see p83).

TOMATOES

In cooler climates, your best bet is to sow seeds from cherry tomatoes, which mature more quickly than larger varieties. This is especially important if you are growing them outdoors. Each ripe fruit contains a large quantity of seeds that you can remove. Put these in a sieve under the tap to wash off the pulp, then dry them on kitchen paper before sowing. For growing instructions, see p148.

PEPPERS

The seeds of virtually any shop-bought pepper (sweet or chilli) can be planted. Peppers need to be started as early as possible, so sow the seeds in late winter. They love sun and heat, so a sunny windowsill is key. If you live in the north of the UK, you are more likely to get a harvest from peppers grown indoors. When flowers appear on peppers that you are growing indoors, gently shake the plants every couple of days to aid self-pollination (see pp150–151 for more details).

PEAS

Sow dried peas in cardboard tubes filled with compost and keep the compost moist. If you are sowing peas for pea shoots, you can do this any time of year. If you are hoping to grow peas to maturity, sow them in mid- to late spring for best results. When seedlings are ready to go outside (see p132), plant the roll and its contents out next to supports – I use old Christmas tree branches to start off, adding bamboo canes or tall sticks as the plants grow. When you harvest, remember to leave some pods on the plant, so that when these have dried, you can extract the peas to plant next year.

CHICKPEAS

Start dried chickpeas indoors in early spring as you would peas (*see above*). Transplant them outdoors three weeks after the last frost, and you may be able to enjoy fresh chickpeas straight from the pod – a rarity in the UK.

TIP

Why not try to grow seed from a heritage tomato rather than from a supermarket tomato. Once you've collected and dried seed from ripe fruit, you will have dozens of packets to take to a seed swap.

Chickpea plants grow to around 45cm (18in) high so finding suitable supports is relatively easy. Chickpeas are far more drought tolerant than garden peas. It takes around 100 days after planting for them to produce a crop. Pinch off the green pods and enjoy the taste of the fresh chickpeas inside – you only get around two per pod. Save some to plant next year.

BEANS

If you find a pack of dried broad beans, also known as fava beans, start them off indoors in early spring, then plant them out when they're around 7cm (2¾in) tall. They usually won't need support.

Growing types of beans common in kitchen cupboards, but uncommon in the garden, will give you fantastic plants to trade in plant swaps. My favourites include kidney, soy, and borlotti. Many gardeners, myself included, are really excited by the prospect of trying out new crops; in such cases, getting a harvest is considered a bonus rather than an expectation. Sow borlotti, kidney, and soy beans indoors from mid-spring and you may be blessed with good harvests from midsummer onwards.

GARLIC

Garlic is easy to grow, with a single clove becoming an entire bulb in the ground. Shop-bought garlic may have been pre-treated, but it is always worth a try. Plant cloves in autumn, and leave over winter. Water well during dry weather to help them mature. Gently pull up plants to harvest bulbs once the lower half of the leaves have browned. After your first harvest, you'll have your own to store and plant out in autumn.

POTATOES

Old potatoes that have started sprouting are perfect for planting. If you want more plants, cut the potato into pieces around each "eye". Each piece should grow into a separate plant. They are so productive that you can even grow them from thick potato peelings provided an eye is present. For more growing instructions, see pp139–141.

Each seed potato will produce a large crop, so by the second year you can easily plant 10 times the number of potatoes you did the previous year for no extra cost. Blight can be an issue – see p166 for more information on how to prevent it.

SOME MORE VEGETABLES TO TRY

- **Coriander** Plant the seeds sold as a spice to grow the feathery leaves. Try sowing mustard seeds, too.
- **Ginger** Cut the knobbly root ginger into pieces, so that each has an eye – the small yellow tips (*see right*). Plant each piece in compost, so the eyes are level with the surface. The green shoots taste wonderful.
- **Pumpkins** Save pumpkin seeds at Halloween. Clean them, then dry them on a windowsill. Plant them the following spring.

1

2

3

TEST YOUR SEEDS' VIABILITY

To work out which of your seeds are viable, take a sample of a specific seed – such as a few dried peas – and do a germination test to see if they sprout. The proportion of seeds that germinate from your sample will give you an idea of how many will germinate from the rest of the packet. This will save you wasting precious compost, as well as time spent planting and watering seeds that won't grow. Test your sample seeds two or three weeks before you plan to sow the rest of the batch.

1. Line the container with a double layer of paper towels (or similar), and moisten with water using your mister. The towels should be wet but not completely sodden. Place your seeds on the paper towel so that they are evenly spaced. Put the lid tightly on the container and place the whole thing in a warm spot, such as on a ledge or windowsill above a radiator, or in an airing cupboard.

2. Open the lid and check the paper towels every two or three days, re-moistening with the mister so that they don't dry out.

3. Within a week of starting your test, you should see roots starting to emerge. If there are no signs of germination after two weeks, then it's unlikely that the seeds are viable. If some of the seeds have formed roots, cut the paper so these are separated from the rest of the group. They should then be planted in compost with the paper still attached, so you don't damage the roots and waste germinated seeds. The paper will soon break down.

YOU WILL NEED

- an old ice-cream tub or plastic container (with lid)
- paper towel, tissues, or paper napkins to double layer the base of the container
- a misting device
- six to 10 seeds to test

SWAPPING SEEDS, MAKING CONTACTS

A seed-swap event is the perfect opportunity to get some seeds without spending any money – it's a truly inspiring example of a local community coming together and making the most of their spare resources.

The idea is simple – bring excess vegetable, flower, and herb seeds, and swap them for seeds you don't have. Swaps are generally carefully organized, with plant types separated into individual boxes.

HOW TO FIND SEED SWAPS

Start by contacting local gardening clubs. You can also check town halls and markets, as well as posts on local websites and Facebook groups. These events usually happen in late winter and early spring.

WHAT TO TAKE WITH YOU

Your first seed swap will be your most important, because you need to find a couple of crops that you can easily grow and save seeds from in your first year. You'll then have seeds to bring to swaps the following year. Peas and runner beans are perfect for this, so you'll be in an amazing position even if you only manage to get these types of seeds.

If you are in your first year of growing vegetables and haven't saved any seeds yet, ask the seed-swap organizers if you can donate some other items instead. I have noticed that many seed swaps feature a few other items, so there's a good chance that your suggestions will be welcome – if you don't ask, you won't find out, and you may miss an opportunity. These might include:

- perennial plants (see pp88–115)
- pots or upcycled containers
- gardening books
- homemade seed labels
- spare tools

An even better bartering tool to offer is simply your time, as some help with looking after the stalls is often welcome. This is also an excellent way of talking directly to other gardeners and asking them for their tips and tricks on how to grow specific plants.

When you have managed to save more seeds to exchange and are confident about how to grow the seeds you are picking up, consider taking a wider variety of seeds with you.

MAKE THE MOST OF SEED SWAPS

Get to the seed swap when it opens to ensure you can get some of the seeds that are harder to save, such as beetroot or kale – these pollinate in the second year, or easily cross-pollinate (see p129 for the easiest vegetables from which to save seed yourself).

If you have more seeds in a packet than you can sow yourself, don't let them go to waste. You can give them to a friend or even start them off as seedlings to take to a plant swap later on (see pp86–87).

SEED-SWAP ETIQUETTE

Here's how to make sure the seed swap is fair for everyone involved:

- **Don't take more seeds than you need** – everyone should have a chance to get a good range of seeds.
- **Pick up a variety of seeds** – I take no more than two or three packets of each type of seed I want, so other people also have a variety to choose from.
- **Take your time** – allow people time to choose, and don't push because you see something you want.

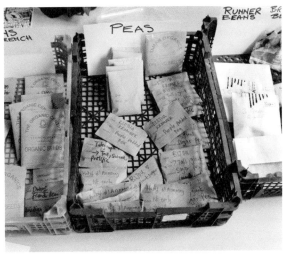

IF YOU DON'T HAVE ANY SEEDS
TO SWAP, YOU COULD OFFER
TO HELP LOOK AFTER ONE
OF THE STALLS INSTEAD

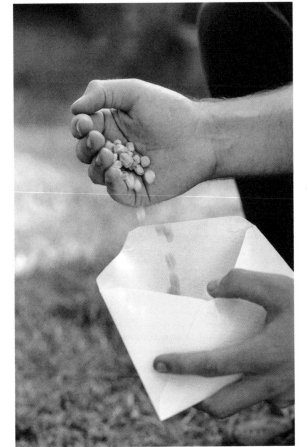

TIP

You don't have to wait for a plant swap to exchange spare plants. Get to know other gardeners through clubs or local allotments and arrange to swap seedlings informally whenever you have plants to spare.

PLANT SWAPS

Many gardeners, myself included, over-sow and have excess plants that are perfect for taking to plant swaps. Note the swap date and, if you have seeds to spare, start sowing around two to four weeks beforehand (*see box below*).

Plant swaps usually happen in late spring, when gardeners take stock of their surplus seedlings and young plants, hoping to find good homes for them. There is plenty of time to grow plants from the seed you picked up at seed swaps (see pp84–85), as these tend to happen in late winter. Organizations or groups that host seed swaps will often host plant swaps, too, so you will already be familiar with the setup. I always find plant swaps exciting, as you often come away with a range of interesting plants that can go straight into your growing area for that satisfying "instant garden" effect.

WHERE TO START

You can grow plants to swap by sowing excess seed from seed packets that contain more seed than you have space for, or by sowing surplus seed collected from your own crops. See the table below for some popular, easy-to-grow plants to take to plant swaps and how to start them off.

WHAT TO LOOK FOR

Choose strong, healthy seedlings when selecting plants at a plant swap. A sad-looking plant may bring unwanted pests and diseases to your garden, or simply die and prove a waste of your time, energy, and growing space. The best plants for you to take will have bright green leaves with no marks or signs of damage, and strong upright stems. It's useful for plants to be labelled so that you know what variety of a vegetable you are growing. That said, I would rather grow an unnamed crop than none at all.

SOUGHT-AFTER ITEMS

People always like to try something out of the ordinary, such as pepper plants grown from shop-bought peppers (make sure you label these honestly) and chickpea seedlings from packets of dried chickpeas. In future years, you'll have the option of propagating cuttings and layers from your own perennial fruit and herbs, which always go down well at swaps.

PLANT	TIMING	SOWING INSTRUCTIONS
Squash, such as courgettes	3 weeks	Sow one seed per pot three weeks prior to the swap so the plants are a decent size.
Tomatoes	3–5 weeks	Start seeds three to five weeks before the swap date and grow one plant per pot.
Brassicas	2–3 weeks	Use a tray to sow 10–15 seeds and start two to three weeks prior to the plant swap.
Lettuce	1–2 weeks	Sow seeds in trays one to two weeks before the swap. Don't thin them out so you have plenty of seedlings that people can transplant themselves.
Broad beans or runner beans	3–4 weeks	Sow one seed into a cardboard roll, six rolls per tray, four weeks prior to the plant swap.
Peas	3–4 weeks	Sow three to four seeds per pot three to four weeks before the plant swap.

HOW TO GROW PERENNIALS

ESTABLISH THESE PLANTS EARLY ON
FOR HARVESTS THE FOLLOWING YEAR,
AND FOR MANY YEARS TO COME

PERENNIAL CROPS

When growing food for free, perennial fruits, herbs, and vegetables are invaluable. Once these hardy plants have put down roots, they require little maintenance and you can depend on them to provide harvests for many years.

The many different types of edible perennials can be divided into herbs, crowns (such as rhubarb and asparagus), tubers (Jerusalem artichokes), creeping berries (strawberries), trailing berries (blackberries and hybrids), and fruit bushes (currants and gooseberries).

Each perennial is grown and propagated differently, but the techniques aren't complicated, and I have deliberately chosen the easiest growing methods to minimize time, effort, and compost requirement. The advice on this page covers routine care and maintenance of perennials, as well as harvesting tips and how to increase your stock through propagation. For more growing instructions for specific types of perennials, see pp92–115.

WEEDING

When harvests of perennials are months away and your energy is primarily focused on quick-yielding annual vegetables, it's easy to forget about weeds. To suppress weeds around perennial fruit bushes, crowns, and climbing fruit, I put a few layers of cardboard (or thick layers of newspaper) on the ground and anchor them in place with stones (or bricks). I then pull out any weeds growing around the stem – a quick and easy task that you need to do only once or, at the most, twice a year.

FEEDING

Perennials aren't nutrient-hungry, so most of my home-made compost is applied to my annual crops. However, when transplanting a perennial into its permanent position, I do add some compost to make it feel welcome. Dig the planting hole twice as deep as the plant's rootball and drop in a few handfuls of compost before planting to encourage the roots to grow downwards. Don't worry if your compost isn't ready, the perennial will still produce a good yield.

In spring, if I have some spare compost left over, I like to spread a 3–5cm (1–2in) layer on top of the cardboard weed-suppressing mulch to give perennials a boost. This isn't strictly necessary, though: brambles, for example, produce abundant crops of delicious berries in the wild without us feeding the soil.

HARVESTING

Unlike many annual vegetables that crop over many months or ripen at different rates, perennials will always be ready to harvest at a certain time of the year. Fresh is best so check your crops every couple of days and enjoy your produce at its peak of ripeness. Berries, such as raspberries or strawberries, must be picked ripe or they will quickly spoil and turn mouldy.

PROPAGATION

Creating new perennial plants is what first stirred up my passion for gardening. As a child, I would stick sections of stem into the ground and be amazed when they took root and turned into new plants. Soon, I had more plants than we had space for, so I started selling them or giving them away to friends and neighbours. For the perennials in this chapter, the three key methods of propagation are: taking cuttings, division, and layering – all of them great ways to increase your stock for free.

Perennials can provide a good harvest over the long term (*top right*). Suppress weeds by placing cardboard around plants (*top left*) and, if you have enough compost, mulch each year (*below left*). Create more plants using simple propagation techniques, such as layering (*below right*).

ONE YEAR, ALMOST ALL OF THE 300 CUTTINGS I TOOK FROM OUR SOFT FRUIT BUSHES ROOTED AND GREW INTO A NEW PLANT

HERBS

In the kitchen, freshly picked herbs will make your home-grown produce taste and smell fantastic. They are very easy to grow and propagate, and some are evergreen, so you can maintain a constant supply for harvesting all through the year.

Think about the herbs you use most in cooking and prioritize growing these. My all-time favourite herbs to grow are:

a. Chives
The grass-like leaves of this popular culinary herb have a mild onion flavour. The pretty pink or purple flowers are also edible. It will cope in partial shade and slightly damp conditions. Chives will die back over winter.

b. Lavender
A popular, fragrant herb often used in desserts. Grow lavender in a sunny spot to attract bees and other beneficial insects to your growing area. Take cuttings (see p96) from non-flowering shoots. The leaves are evergreen, but the flowers are only present in summer.

c. Lemon balm
This aromatic herb has sweetish, lemon-scented leaves that are used to make calming teas. It is best grown in a container to confine its wandering roots. Lemon balm dies back over winter.

d. Lemon verbena
The strongly citrus-scented leaves make a fantastic lemony tea; the flowers have a milder lemon taste. Grow in well-drained soil and position in full sun. Lemon verbena will lose its leaves over winter.

e. Lovage
Grown for its celery-flavoured young leaves, lovage is a sizeable plant and may reach 2m (6½ft) in height. It prefers rich, moist soil and full sun or partial shade. Lovage dies back over winter.

f. Marjoram
Used in a variety of Italian dishes, marjoram has a mild spicy flavour. Grow in a sunny spot with well-drained soil. Marjoram dies back over winter.

g. Mint
Chopped, fresh mint leaves are delicious with new potatoes and peas. Like lemon balm, mint is best grown in a container to restrict its invasive roots. Mint will die back over winter.

h. Rosemary
A shrubby herb, rosemary can be harvested fresh throughout the winter and is easy to propagate from cuttings (see p96). I love to add sprigs of fresh rosemary when frying potato slices in olive oil.

i. Sage
This versatile herb has been cultivated for centuries, and can be harvested throughout the year. The young leaves are delicious fried in butter. Sage dislikes damp, so grow it in well-drained soil in full sun.

j. Tarragon
The pointed aromatic leaves have a fantastic aniseed flavour. In colder areas, protect the crown from frost. The leaves will die back over winter, when propagation from cuttings is less successful, but still possible.

k. Thyme
A sun-loving herb, thyme is widely used for cooking, and bees love the summer flowers. Grow in well-drained soil and harvest throughout the year.

START OFF

All of the perennial herbs mentioned in this section can be grown in large containers or tyres lined with old compost bags. I like to use tyres because they contain any spreading roots and allow a few different herbs to be grown together. See pp32–33 for instructions on planting up tyres. Herbs prefer good drainage, so when using tyres or other containers, I always add a handful of stones or broken pottery at the base. Fill the tyre with an 80/20 mix of topsoil and compost.

GROW ON

For deciduous herbs that die back over winter, cut back the old growth in late autumn. Keep the plants clear of weeds and in winter apply a 2–3cm (¾–1¼in) mulch of compost around the base. In colder parts of the country, insulate rosemary, thyme, and the crowns of tarragon with straw (or an old blanket) during prolonged frost or snow. Lemon verbena is best grown in pots and brought under cover over winter.

Once the growing season starts in late spring, feed herbs in containers once every two to three months with homemade liquid feed (see pp72–75).

ATTRACT POLLINATORS

Growing a range of flowering herbs is one of the best ways of encouraging beneficial insects to visit and pollinate your fruit and vegetable crops. Marjoram, lavender, and chives are my top three choices for enticing bumblebees, honeybees, hoverflies, and butterflies into my garden.

HARVEST

Herbs have a very long cropping season and evergreen types can even be picked through the winter. Harvesting herbs is as simple as taking what you need and leaving at least one half of the plant untouched to maintain strong growth year on year. At the rate herbs grow, it is hard to overpick.

PROPAGATE

There are two principal methods for propagating established perennial herbs: division and taking cuttings. It is possible to grow perennial herbs from seed, but they are hard to source and take much longer to grow and establish than the two main propagation methods.

How to propagate by division

After carefully digging up a mature plant (or removing one from its container) in early spring or autumn, slice through the roots to create up to four divisions.

Top herbs to grow from divisions:
- chives
- lemon balm
- lovage
- marjoram
- mint
- thyme

1. Uproot the herb by digging beneath it and using a fork to lift the plant out of the ground. Shake off excess soil, and set the plant down on a lawn or soft ground.

2. Use a spade to make divisions, pushing it firmly down the middle to cut the plant in half. This may seem ruthless, but the herb will quickly recover from any damage. When splitting a large clump, divide the halves again to make a maximum of four divisions.

3. You now have two to four mature herbs, each with a good root system. Plant each into their final positions in the ground or in a container – I've used a tyre. Potted herb divisions also make fantastic gifts for neighbours.

When taking rosemary cuttings, prioritize shoots that have not flowered.

How to propagate by cuttings

My favourite method by far – I can have dozens of rooted herbs to transplant or swap in weeks! Take cuttings from late spring through to midsummer, so they have a chance to develop strong roots before winter. Not all the cuttings in a batch will take root.

Top herbs to grow from cuttings:
- lavender
- lemon balm
- lemon verbena
- marjoram
- mint
- rosemary
- sage
- thyme
- tarragon

You will need a sharp pair of scissors or a sharp knife, some empty yogurt pots (with drainage holes) or plant pots, and enough homemade compost or a 50/50 mix of compost and topsoil to fill the pots.

1

2

1. Cut sections of stem about 7–10cm (2¾–4in) long from the growing tips of the plant and put your cuttings in a jar or container of water.

2. Fill the pots with the compost or compost/soil mix to just below the rim.

3. Use your fingers to strip the bottom two-thirds of leaf from the stem

4. Place the cutting into the pot and push it down, so the bottom set of leaves sits above the surface of the compost. Water. Leave the cuttings in a warm, sunny location and make sure they don't dry out completely. Never overwater cuttings or the stems will rot.

Help cuttings reach maturity

After six to eight weeks when the cuttings have rooted, fresh foliage will start to appear. Transfer cuttings in yogurt pots to slightly larger pots after two months; leave those in plant pots for three months before potting on. I leave my potted herb cuttings

outside in a sheltered place over winter, but keep cuttings of less-hardy lemon verbena, rosemary, and tarragon indoors on a cool windowsill. In late spring, once growth has started, they can be planted in their final positions. With shrubby herbs, pinch out the tips of the cuttings when they are around 10cm (4in) tall to encourage branching and a good shape.

TIP

Try rooting a stem of mint in water. Strip off the bottom two-thirds of leaf and place in a jar of water. Two weeks after the first roots appear, take out the stem and pot it up in some compost.

3

4

RHUBARB

Strictly speaking, rhubarb is classed as a vegetable (rather than a fruit) because we eat the stems. It grows from a rootstock or "crown", is easy to grow and propagate, is very low-maintenance, and can easily live for more than 10 years.

Rhubarb can be grown from seed, but planting a piece of the crown gives much quicker results and divisions are easier to source than seed. It is a fantastic crop to harvest during the so-called hungry gap – the period in spring when there is very little fresh produce around.

START OFF

In winter or early spring, plant the rhubarb division directly in the ground or in a good-sized raised bed. Rhubarb has large stems and leaves and it's best to start with just one plant until you get an idea of yield. Make the planting hole twice as deep and twice as wide as the division, and add compost or well-rotted manure to the base. Plant so the tip of the crown is just poking out of the soil, then refill around the sides with soil from the hole and water well.

GROW ON

To keep weeds from suffocating the crown, arrange a small collar of torn-up newspaper or cardboard around the stems. Every winter, I like to mulch around the crown with a few centimetres (an inch or two) of compost, manure, or partly composted material to give my rhubarb plants a boost. Water in summer, unless rhubarb is growing in the shade, and divide the crown every five years to reinvigorate the plant and maintain good yields (see propogate, *opposite*).

1

2

HARVEST

Start harvesting rhubarb stems from early spring, and enjoy crops through to midsummer. If your rhubarb was planted as a division, take no more than a quarter of the stems during the first year. From a mature plant (more than three years old), take up to a half of the stems at a time, but make sure the plant sends up new stems before picking any more.

PROPAGATE

Divide rhubarb in mid- to late autumn when the leaves have died down, or in early spring just as the temperature begins to rise.

1. Carefully loosen the ground around the mature crown with a fork then gently lift out the whole plant. Set it down on a piece of soft ground.

2. Once the rhubarb crown is uncovered you will be able to see the buds. These will produce the stems. Use a spade to cut the crown into sections, each with at least one bud and some roots. If the roots are very tough you may need to be a bit more forceful with the spade.

3. Immediately transplant the divisions into their final position so they don't dry out. Rhubarb is a popular plant, so if you are planning on using the divisions as bartering items, pot them up and exchange for another fruit or vegetable plant.

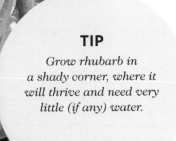

TIP

Grow rhubarb in a shady corner, where it will thrive and need very little (if any) water.

ASPARAGUS

This perennial crown, like rhubarb, is also ready for harvesting during the hungry gap. I haven't included growing instructions for asparagus in this book because it is much harder to source and grow than rhubarb. That said, it is definitely a plant you should consider for the future.

JERUSALEM ARTICHOKES

Apart from potatoes, which are treated as annuals, my favourite perennial tubers are Jerusalem artichokes. Above ground, their stems can reach up to 3m (10ft) and produce pretty yellow flowers.

Plant swaps are a good source of Jerusalem artichokes, which are grown and harvested like potatoes. Unlike potatoes, however, they don't suffer from the diseases that mean you need to replant them again each year – our Jerusalem artichokes have been left in place for over 15 years since we first planted them.

The digestive issues associated with Jerusalem artichokes (and, in case you were wondering, you build up a tolerance fairly quickly) are significantly outweighed by its many advantages. They grow well in partial shade, require very little watering, crop through winter until early spring, are relatively free from pests and diseases, and are extremely low maintenance – they tick all of the boxes.

START OFF

These vigorous relatives of the sunflower are best grown in raised beds or in the ground.

1. In early spring, dig a trench approximately 25cm (10in) deep. Cover the base with a 5cm (2in) layer of vegetable peelings or well-rotted manure followed immediately by a light layer of soil, so you don't overwhelm the tubers with nutrients right away.

2. Place tubers at 30cm (12in) intervals along the trench, on top of the light layer of soil.

3. Backfill the trench with the soil you removed. Leave 30–40cm (12–16in) between rows.

1 2

GROW ON

Shoots appear in mid-spring and grow at speed – even faster than weeds. The shade cast by the tall, leafy artichoke stems also keeps the ground around them clear of weeds. In dry summers, water once a week to keep the plants healthy, and mulch the base of the stems with grass clippings to help retain soil moisture. Jerusalem artichokes grown in a sunny location will reward you with lovely sunflower-like blooms around the start of autumn. Don't cut down the stems when they start to die back; leave them standing to feed the tubers developing below ground.

HARVEST

Start harvesting the artichokes from late autumn onwards, either pull each plant up by its stem, or use a fork to lift the tubers. I'm always amazed by the size of the harvest from just one plant. You can also lift the tubers as and when you need them; they won't come to any harm if left in the ground over winter.

PROPAGATE

Jerusalem artichokes really couldn't be easier to propagate. As you harvest a row, simply leave one tuber in the ground at the same depth and at the same spacing you planted it (20cm/8in deep every 30cm/12in). Then add two to three handfuls of compost to meet each tuber's future growing needs and push the soil back to cover the row. A planting of three initial tubers could easily increase to 20 tubers the following year, so starting small and not eating your first harvest will give you incredible yields from the second year onwards.

TIP

For the best results, give Jerusalem artichokes – the tallest plants in this book – a sheltered spot to grow in.

BLACKBERRIES AND HYBRIDS

Cultivated forms of this trailing plant and its hybrids – tayberries and loganberries – produce lots of large, succulent fruits over a long period. Treating them as climbers is a great way of growing your own berries in a small space.

Blackberries are easy to grow and just as easy to propagate. There are many varieties to choose from – my favourite is called 'Silvan' or silvanberry – and some are bred without thorns, so you won't need to wear gloves when working with them. Be aware that a new plant won't fruit until the following summer, but your first harvest will be worth the wait. Hybrids such as the tayberry, produced by crossing a blackberry with a raspberry, are grown and propagated just like blackberries and yield fantastic crops.

a. Blackberries
Wild brambles are rampant and very thorny, but the specially bred varieties (seen here) have bigger fruits and can be kept under control when grown against a support.

b. Silvanberry
This variety produces heavy yields and the dark-red ripe fruit are deliciously sweet – a lovely garden snack.

c. Tayberries
These berries are similar to loganberries, but sweeter and larger in size. Choose tayberries if you prefer not to add sugar to your berries.

d. Loganberries
Sharper-tasting than tayberries, these fruits are a wonderful shade of red. You can eat them fresh, like raspberries, or turn them into an amazing jam.

a

b

c

d

START OFF

Blackberries grow well in the ground or in a large container. Train them along a fence, against a trellis, or up a sturdy support. You can also grow them against a wall as long as there are wires to tie the stems onto.

When planting blackberries, always choose a warm, sunny spot – they need warmth at the fruiting stage to ripen well. Container-grown plants can be planted out at any time of the year, but bare-root plants must be put in the ground during winter, when they are dormant. Soak bare-root plants for two hours before planting out.

1. Dig a hole wide and deep enough to accommodate the roots without squashing them, then dig down another 5cm (2in). Make the hole as close to the support as possible, so that you don't have to bend the stems in order to train them.

2. Add a layer of compost about 5cm (2in) deep to the base of the hole and water it thoroughly.

3. Next place the plant in the hole, making sure the soil is at the same level it was in the container. With bare-root plants, check the base of the stem for a line where the colour changes from dark to light – this is where the soil level should be in their new site.

4. Once you have positioned your plant correctly, refill around it with the soil you dug out. Add an extra handful of compost at this stage to welcome the plant into its new home.

5. After filling the hole, use your feet to compress the soil on either side to firm the plant in, and then give it plenty of water.

TIP

We have a variety of berries growing on a fence along one side of the garden – they provide welcome wind protection as well as a great crop of fruit.

GROW ON

Ensure that the stem is kept clear of weeds. This may only require one weeding session in spring and one in summer, but keep on top of it.

In autumn, I like to give my berries a 5cm (2in) mulch of compost, covering a radius of 20cm (8in) around the stems. Mulching is particularly important for topping up nutrient levels if you're growing plants in containers. Cover the ground with cardboard before mulching to suppress weeds.

Train new growth

In late autumn, cut out all the stems that fruited this year then separate out the new growth, which will be greener in colour, and tie these stems to your support. Blackberries and their hybrids all fruit on the previous year's growth, so train the new shoots, which you will be picking fruit from next year, in one direction, so that you can train next year's shoots in the opposite direction. Tie each stem in three to five places, starting at the base and working up, to keep it nicely secured, and space the different stems about 10cm (4in) apart in a fan shape.

When you run out of space on your support, keep any spare stems to propagate (*see opposite*) or cut them off and add them to your compost heap.

Tie new shoots onto your support at intervals in autumn, after you have cut down the old growth that fruited in summer.

HARVEST

Berries give fantastic crops most years, but the fruits don't last long once harvested – perhaps three days at most in the fridge. Eat them at their freshest and sweetest or freeze them to use at a later date. I prefer to harvest them towards the end of a sunny day to allow some extra time for the berries to sweeten.

Garden birds will be taking an interest in your berries, so it's worth placing some homemade bird-scarers (see p164) around the plants when the fruits start to ripen. Otherwise you may find that they are all gone before you can get to them. After harvesting your berries, cut the stems that bore fruit down to the ground in autumn.

1

2

PROPAGATE

Propagating blackberries and their hybrids by "layering" – burying the tips of new shoots that haven't fruited – is very easy. Just be sure not to cut off the stem that is attached to the "mother" plant too early, or let the container dry out for too long. Layering is best done from late winter to mid-spring, as this allows your plants the maximum time to develop over the growing season. When your chosen berry has rooted and become mature, you can propagate up to a dozen new plants from it each year – a great resource for bartering.

1. Choose a medium-sized container such as an old bucket. Make drainage holes in the base then fill with compost or a mix of compost and topsoil.

2. Select a long stem from new growth that did not bear any fruit. Place the container on the ground in a position where you can easily bend the stem down to meet the surface of the compost.

3. Bury the tip of the stem a few centimetres (about an inch) into the compost at the centre of the container and use a small stone or a forked twig to hold it in place.

4. Rainfall should keep the rooting stem moist over the winter, but you may need to water once a week during dry weather in spring and summer. A shoot will appear in early spring and then grow on rapidly. Keep the original stem attached to this plant until the end of summer, then use secateurs to cut it off. Your new plant will be ready to plant out in winter and will bear a small crop the following year, followed by an abundance of fruit the year after.

KEEP HUNGRY BIRDS AWAY FROM YOUR RIPE BERRIES WITH SIMPLE HOMEMADE SCARERS

3

4

STRAWBERRIES

These delicious, iconic berries conjure up warm summer days and bring
so much joy. There is nothing quite like the taste of the first ripe, home-grown
strawberry that you've picked and eaten straight from the plant.

Strawberries are easy to grow in beds or containers and they don't like soil that is too fertile. This makes them the ideal crop if you are new to growing food and don't yet have a good supply of your own compost to enrich your soil. Most varieties – apart from perpetual strawberries – have a single cropping period every year, and the best ones to start off with are the summer bearers.

START OFF
Strawberries are the easiest plants in this book to propagate, which means you won't have trouble sourcing them. I find young strawberry plants are readily available at plant swaps, usually in spring. Plant them 30cm (12in) apart in containers or raised beds filled with topsoil, but you can add a handful of compost or bury some vegetable scraps 10cm (4in) below the surface to get the plants off to a good start. Spring is a good time for planting out, or alternatively from mid-autumn through to winter. You can expect harvests from the first year of planting, so if you obtained some strawberries in spring, you can look forward to eating some in summer. These creeping perennials send out long shoots, called runners, which can easily take over any spare ground. It's wise to contain strawberry plants by growing them in in a tyre planter or a raised bed with edges.

GROW ON
Apply a light 2–3cm (¾–1¼in) layer of compost around the plants every spring or two. When strawberries begin fruiting, it's a good idea to surround the plants with dry grass clippings or straw to raise the fruits off the soil and keep them in tip-top condition. If the temperatures are forecast to dip below freezing and flowers have formed, make sure you protect the plants from frost (see p24). Sever any runners from strawberries that are flowering or fruiting because these divert a lot of energy from the plant.

HARVEST
Pick strawberries when the whole fruit has turned red. I love harvesting strawberries around midday when the sun has warmed the fruits – it seems to intensify the sweetness. Garden birds will also be keeping a close watch on your ripening fruits, so take preventative measures (see p164) and you will avoid extreme disappointment. Strawberries spoil very quickly after harvesting, so eat them within two days, which shouldn't pose a problem! If you can dedicate a whole bed to the fruit, you can turn your crop into fantastic preserves.

ALPINE STRAWBERRIES
Some varieties have runners and some don't, but these miniature strawberries couldn't be easier to grow. Just feed them once with a light mulch (2.5cm/1in) of compost and leave them to it. You can grow alpine strawberries from seed and enjoy the tiny but extremely flavourful fruits, but the harvests won't make much of a dent in your appetite.

Surround your strawberries with a mulch of dry straw to retain moisture and keep fruits off the soil (*above*). Alpine varieties are delicate and delicious (*top right*). Regular strawberries (*centre and right*) have bigger flowers and larger fruits.

PROPAGATE

Although they are perennial, strawberry plants do lose their vigour, so replace strawberry plants every three to four years with newly propagated plants to maintain good yields. This can be as simple as letting some runners root into the bed and then removing the "mother" plant in autumn. Strawberries, like blackberries (see pp102–105), can be propagated by layering. When the runners sent out by the plants trail along the ground, the growing tip soon starts to send out roots. Once the roots have made contact with the ground, a strong individual plantlet, known as a layer, will develop. In time, this layer will send out its own runners. The process is simple, and in common with most gardeners, I'm always happy to pass on surplus plants.

Propagating from layers

I usually start propagating strawberries from mid- to late summer, once they have finished fruiting and have channelled their energy into creating new plants by layering. All you need are some small- to medium-sized pots, an 80/20 mix of topsoil and compost, secateurs or a sharp knife, and a few small stones to act as weights.

1. Choose as many of the layers (the tiny plantlets growing at the end of the long, thin runners) as you want and remove the rest. The mother plant will now focus its energy into these "daughter" plants, which will quickly take root. Some layers may already be showing roots before they have even made contact with the ground.

2. Fill the plant pots with the mix of topsoil and compost, then take a chosen layer and carefully place it in the centre of a pot with the runner still attached. Firm in some compost around the base of the plant. Repeat with your other chosen layers.

3. To ensure each layer is secure and to encourage it to form roots, weight each of them down with a small stone. Place the stone in the pot next to the layer and on top of a section of the runner that attaches it to the mother plant. You could also use a piece of bent wire to keep the runner and layer pegged down. Sometimes a layer will develop a secondary runner. If that happens, simply cut it off.

4. If the pots you planted the layers into are sitting on grass, slide a sheet of cardboard underneath them to stop weeds growing up around them. Keep the layers watered, although at this stage they won't need too much watering as they will still be receiving moisture and nutrients from the mother plant via the runner.

5. Keep the runners attached to the mother plant for a couple of months, then cut them free. Strawberries are hardy and can be left outside over winter. In spring, the individual plants will send up fresh growth. You can either transplant them or use them as bartering items.

STRAWBERRIES ARE NOT ONLY DELICIOUS BUT ALSO THE EASIEST PLANTS TO PROPAGATE IN THIS BOOK

FRUIT BUSHES

When I first started growing soft fruit and saw how easy it was to increase my stock, I became even more passionate about producing food. Whether you go for currants or berries, these bushes are low-maintenance and just keep on giving.

To get the best crops from perennial soft fruit bushes, plant them directly in the ground. They will also grow well in large containers, filled with a 80/20 mix of compost and topsoil, and deep raised beds. Given fruit bushes are so easy to propagate, you shouldn't have any trouble sourcing them at plant swaps, and once you have established plants, you can easily create more. Perennial fruit bushes should be planted out only when the foliage has died back and they are dormant (from late autumn to early spring) – as long as the ground isn't frozen or waterlogged. When planting, space the bushes 1.2–1.5m (4–5ft) apart. They prefer a sunny location, although currant bushes also produce respectable yields in partial shade.

BLACKCURRANTS

Blackcurrants crop reliably and, in my experience, suffer less from pests compared with other soft fruits. Their close relatives, jostaberries, are the result of a cross between a blackcurrant and two types of gooseberry. Jostaberry fruits are somewhere between blackcurrants and gooseberries in size, but I think they taste more like a blackcurrant when ripe.

Below, I outline everything you need to know to grow blackcurrants successfully, and the techniques given can also be used to grow jostaberries. If you look after your blackcurrants and jostaberries well, you can expect them to live for over two decades.

START OFF

Blackcurrants send up many shoots from the base and look a bit like shrubs that have been coppiced (pruned to the ground to encourage growth). To help multiple stems to grow from young plants, plant them deep in the ground, so that the crown is well below the level of the soil. Plant out blackcurrants during dormancy, when the stems are bare.

1. Dig a hole around twice the size of the rootball of your blackcurrant plant and position it in the hole. Check that the forked branches above the stem are at least 2.5cm (1in) below the soil level. If they aren't, make the hole deeper.

2. When the plant is sitting at the correct depth, take it out of the hole and add two large handfuls of compost or well-rotted manure to the base. Water the base of the planting hole if dry.

3. Put the plant back into the hole, checking the depth again, and use your hands to backfill around the sides of the rootball, using the soil you dug out.

4. When you've filled the hole, water only if the ground is dry. I like to tread carefully around the base of the plant to help firm it in.

5. Apply a mulch of cardboard, held down with a few stones or some woodchip, around the base of the plant to stop weeds competing with the shoots when they emerge.

6. Finally, cut all the stems back to around 2.5cm (1in) above the ground to encourage a flush of fresh, new growth when spring arrives.

GROW ON

Keep the stems clear of grass and weeds by pulling away unwanted growth. If the mulch around the stems has broken down, renew with torn-up newspaper or cardboard. Spread this around the stems so it extends 30cm (12in) out from the base of the plant. A mulch of wood ash in early spring will improve crop quality. You can save the ash from a wood stove or wood fire. Every other spring, I also add 3–5cm (1¼–2in) of compost to encourage strong growth.

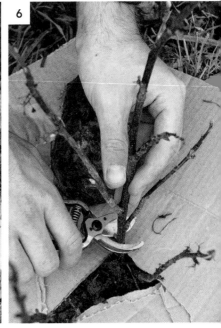

Pruning

Annual pruning of blackcurrants is very easy and will keep well-established plants producing high yields. Every winter, cut a third of the oldest (thickest) stems down to the ground, and remove any stems that are damaged or are growing horizontally. Also look for stems that are crossing, but cut only the oldest of these pairs down to ground level. You are aiming to leave two-thirds of the original stems in place, so don't overdo the pruning!

PROPAGATE

Blackcurrant cuttings planted in pots or in the ground have a success rate of around 90 to 95 per cent. The offcuts from pruning make excellent cuttings, so there is no waste. Pruned material from a single mature blackcurrant will easily yield 20 to 30 new plants.

- After pruning the bush, trim the offcuts into lengths of around 25cm (10in), ensuring that there is a bud at the tip of every cutting. Then make an angled cut 1cm (½in) above (and sloping away from) the top bud, so that rain runs off it.
- Place the cuttings 7–10cm (2¾–4in) apart and at a depth of around 15cm (6in) in pots filled with a 50/50 mix of soil and compost, or in raised beds.
- Leave the cuttings to grow on for a whole year and remember to keep the stems clear of any competing weeds as they grow.
- By the autumn of the following year, the cuttings will have rooted and put on new growth. Once the leaves have fallen off, lift out your cuttings using a fork. At this point you can replant them 30cm (12in) apart to grow on, or put each cutting into a large container filled with an 80/20 mix of soil and compost. If you have several cuttings, put a few into a container with enough compost to cover the roots. Transplant the cuttings using the steps on the previous page, remembering to cut all the shoots down to 2.5cm (1in) above soil level for a flourish of growth in spring. These bare-root cuttings make good bartering material at plant swaps.

REDCURRANTS, WHITECURRANTS, AND GOOSEBERRIES

Redcurrants and whitecurrants look very similar to blackcurrants, but they are grown in the same way as gooseberries, which is why I've grouped them together. These three soft fruit bushes are unlike blackcurrants, which send up many stems, because they grow best on a single stem that forms branches.

START OFF

Late winter and early spring, when the young bushes are dormant, are the best times to plant them out. Use the same method of transplanting as shown earlier for blackcurrants (see p111), but make sure that there is at least 5–7cm (2–2¾in) of each stem above ground before the point where it starts to branch. Also, always plant at the same depth at which the plant was growing previously – look closely at the stem and you'll see a mark where it changes colour from dark wood (below ground) to lighter wood (above ground); plant the bush so that the soil is level with this mark. Plants that have four to five main stems should have all these pruned back so each is about 15cm (6in) in length.

Ripe soft fruits look like mini-jewels and taste delicious. Blackcurrants (*top left*) are grown differently from redcurrants (*top right*), whitecurrants (*below left*) and gooseberries (*below right*).

GROW ON

In early spring, mulch all perennial fruit bushes with a thick layer of potash-rich wood ash (from a wood stove or wood fire) to encourage strong yields.

Keeping these single-stemmed bushes clear of grass and weeds is easier than clearing around the many blackcurrant stems. A mulch works well – I like to use woodchip or compost.

Redcurrants seem to attract more birds, especially blackbirds, than ripe blackcurrants. I tend to harvest them as soon as they turn red and let the birds have a few, but if you have a serious problem, see p164 for tips on deterring birds.

Pruning

Every winter, remove any dead, dying, and damaged stems from your mature plants. Then cut back the new growth, which is softer than mature wood and greenish-looking at the tips, to half its length. Always cut back to a bud that is facing outwards. To encourage a "goblet" shape and allow plenty of air to circulate, cut out any shoots that are growing inwards or cluttering the centre. Think of it as giving the bush breathing space.

SAVE PRUNINGS FROM SOFT FRUIT BUSHES FOR CUTTINGS SO YOU DON'T WASTE THEIR FOOD-PRODUCING OR BARTERING POTENTIAL

BLUEBERRIES AND RASPBERRIES

Compost made from a diverse range of ingredients, as I recommend, will naturally have a neutral pH. This will suit all the other crops mentioned in this book, but not blueberries, which need an acidic soil. Making acidic compost for just one crop is not a good use of time when you are growing food for free, so I haven't included growing information for blueberries. Raspberries have also been excluded because they take up a lot of space in relation to the quantity of fruit you get, and are also less reliable than other soft fruits.

Mulch your currant and gooseberry bushes each spring with a layer of compost (as above) or woodchip.

1

PROPAGATE

You can propagate redcurrants, whitecurrants, and gooseberries using a very similar method to blackcurrants. Use prunings from when you cut back new growth during winter.

1. Set aside prunings that are around 25–30cm (10–12in) in length, making sure there is a bud at the top of each cutting.

2. Keep the top three to four buds on the cutting, but remove any remaining buds by rubbing them off with your fingers.

3. Plant the cuttings in the ground at a depth of around 10cm (4in) and 7–10cm (2¾–4in) apart, or in pots filled with an 80/20 mix of soil and compost. Leave a gap of at least 5–7cm (2–2¾in) between the soil level and the lowest bud. This will encourage the single stem to develop.

4. Leave the cuttings in the ground until late autumn (almost a year after the cutting was first taken), removing any weeds that appear, then lift them. Transplant each cutting into its final growing position or put in a large container filled with an 80/20 mix of soil and compost. Remember to plant the cutting at the same level it was growing at previously.

2

3

4

HOW TO GROW ANNUALS

ALTHOUGH THEY NEED TO BE RE-SOWN
EVERY YEAR, THE DIVERSITY,
PRODUCTIVITY, AND FLAVOUR OF
ANNUAL VEG IS INCREDIBLE.

HOW TO SOW

On the following pages, you'll discover techniques for growing annual crops successfully. First up is seed sowing – an easy, but important skill when growing plants that need to be started afresh each year.

Most allotment and garden vegetables are classed as annuals, which means you sow them from scratch each year. There are several things to consider when sowing seeds, which I cover below. More detail is given within the specific growing instructions for each crop later in the chapter (see pp130–155).

SEED VIABILITY

The benefit of annuals is that they provide an abundance of seed, which makes it easy to obtain until you can save your own. Check that your seeds are viable by doing my simple test (see p83). The seeds of some vegetables remain viable for longer than others (viability length even varies from supplier to supplier with shop-bought seeds). Sowing by the maximum or "use within" storage times given for each crop on the following pages will provide decent germination results. After these maximum storage periods, seeds are unlikely to germinate at a reliable rate, and ultimately this drops so low that the seeds are unusable. Keep your saved seeds in envelopes in a cool dark place for maximum longevity.

SOWING TIMES

To ensure the best chance of a large harvest, it's often recommended that you sow certain crops at a particular time of year. This information is always on the back of seed packets, but I think you can take it with a little pinch of salt. Some crops can be started off much earlier indoors or under protection; others can be sown later for slightly smaller harvests. Certain vegetables are best sown at a particular time, such as parsnips in early to mid-spring, but you can be more flexible with others, such as beetroot. Check the information later in the chapter and find the sowing time for each particular crop.

SOWING DEPTH

You don't really need to get out your tape measure when sowing vegetables. Seeds will still germinate, even if you sow them a little too deep or shallow.

MODULES OR DIRECT?

Annuals can be sown in small pots or modules (including improvised versions, such as cardboard tubes), or directly in the ground. Start seeds sown in modules earlier in the year – use your finger to make a hole in the compost, drop the seed in, and fill in with soil or compost. Keep your seedlings on a windowsill, so that they are protected from pests and the cold. Seedlings will require transplanting outside when they are large enough. Sowing direct is best for seeds that don't mind the cold or don't transplant well.

When sowing direct, press a bamboo cane (see p46) or the top of a rake into the ground to make a shallow drill (use string pulled taut between two sticks to mark a straight line). I then sprinkle seeds into this. Finish by covering the seeds with soil or compost.

SEED COMPOST

Home-made compost is my first choice for seed sowing, but you can mix it with 70–80 per cent soil if your compost supply is limited. Pick out any pieces of material the size of a small coin before sowing, leaving anything that is smaller. In my view, sieving compost when sowing tiny seeds, such as lettuce, is also unnecessary. Think about nature: seeds never fall onto a carpet of perfectly sieved compost.

Sow seeds directly into a shallow drill – here made with the top of a rake and stringline – in the ground (*top*), in modules (*below left*), or in improvised pots, such as these cardboard tubes (*below right*).

HOW AND WHEN TO WATER

Annual plants need a lot of water early on when they are growing strongly,
but too much can cause problems. Learning how and when to water
will ensure your seeds and plants get off to a good start.

You get the best germination rates from seeds if you keep the compost they are sowed in moist. For seeds in pots, modules, trays, and drills, I dip my finger 1cm (½in) into the compost and if it feels dry, I water using a small can or bottle with a fine spray. You may find you need to water daily on warm, sunny days, but only once every three days when the skies are cloudy.

Buckets, pots, and tyre planters dry out faster than raised beds, so water plants growing in these types of container first during a dry spell.

SEEDLINGS
Young seedlings are vulnerable when soil dries out because they haven't yet established strong root systems to take up water from deep in the soil. The first week is the most important, so aim to water deeply at least once every other day during hot, dry weather, but less often as they grow larger. A good time to water is when the top 2cm (1in) of your soil or compost is dry. When seedlings begin to wilt, don't panic. Water deeply and they will quickly recover.

NEWLY TRANSPLANTED
Water seedlings that you have recently transplanted into containers or beds as soon as possible to encourage their roots to grow and spread out. When I transplant seedlings outside, I try to time it just before or during heavy rainfall to save on water – although this may not be possible if you live in a dry area.

YOUNG AND MATURE PLANTS
Once your annuals have more than five or six true leaves (which means a good root system), you can really cut back on watering. For most plants in raised beds or in very large, deep containers, twice a week will be

enough. When there has been some rainfall, you may not need to water mature plants for a month. Where I live, there was a prolonged drought during spring and summer 2018 and some of my mature plants went for two months without water. They not only survived but gave us decent crops. See p27 for tips during droughts.

BEST TIME OF DAY
In summer, the most efficient time to water is before 10am because the moisture has a chance to travel down to the roots of the plant. At midday, most water will quickly evaporate in strong sunlight before it reaches the roots. Any drops left on the leaves can act as mini-magnifying glasses and scorch the foliage. Avoid watering in the evening, as it encourages slugs.

DAMPING OFF
This fungal disease causes seedlings to collapse and wilt. It commonly affects seedlings indoors and is most often caused by poor air circulation, overwatering, and overcrowding – all of which are avoidable. I follow these three easy steps to avoid damping off:
- Sow thinly to prevent overcrowding and choose a sunny spot near a door or window to allow fresh air to circulate.
- Leave small spaces between pots and trays to create good airflow and reduce humidity around the seedlings.
- Water only when the top 1cm (½in) of the compost feels dry. Saturated compost and increased humidity quickly lead to damping off.

I use my improvised watering can (see pp48–49) to water the more sturdy plants in my raised beds.

HOW TO TRANSPLANT

Follow these techniques when transplanting seedlings or young plants
to a temporary or permanent position, especially if you are planning
on growing as much as possible in your outdoor space.

Transplanting is simple as long as you remember to be gentle with seedlings and young plants and keep them watered. There are three key stages:

• **Pricking out** means transferring small seedlings, such as lettuce, from seed trays into individual pots to develop further before transplanting them outside.

• **Transplanting** is moving seedlings with a healthy rootball held together by compost, such as French beans, into their final position outside.

• **Lifting** involves taking seedlings without a rootball, such as leeks, from a seedbed and planting them into their permanent home.

HARDENING OFF

Some tender vegetable seedlings started off indoors benefit from being hardened off before transplanting. Take the plants outside during the day then bring them back inside late in the evening for three or four consecutive days. The seedlings will gradually become acclimatized to fluctuating outdoor conditions. Theoretically, it will be less of a shock for the plants when they are finally planted outside, but in my experience this time-consuming method is only worth doing for more tender plants, such as dwarf French beans, runner beans, and squash. Seedlings need warmth to settle in and grow, so it's a good idea to keep an eye on the weather. Hold back on transplanting if cool, wet weather is forecast.

PLANTING OUT SEEDLINGS

Most of your transplanting will be from pots and trays into containers and beds. Follow these simple steps for success:

1. First, ensure that your seedlings are well watered. Decide where you want to plant them and make small holes with your fingers to mark the positions.

2. If using biodegradable pots, skip this step; if not, turn the container upside down and place your fingers over the top. Tap the base firmly to release the roots.

3. Holding the plant (or the plant in its biodegradable pot) in one hand, use your other hand or a trowel to make a hole around the same size as the roots.

4. Put the plant (including biodegradable pot) into the hole and use your fingers to push soil back around the roots. Gently firm the seedling in place.

5. Continue transplanting the rest of the seedlings and water them well – if using a milk carton, pour water onto the soil immediately around the seedling so it goes straight to the roots.

SETTLING IN

Seedlings can sometimes look a little unhappy in their new surroundings. Be patient. They are working hard to form new roots and will take around a week to settle before putting energy into new growth. During this "settling in" phase, make sure you water them at least three days after transplanting (unless it rains heavily).

Transplants without a developed rootball, such as brassicas and leeks, are particularly vulnerable after being planted out because some roots will have been lost when they were lifted. Keep them well watered and they will soon recover.

PROTECTING SEEDLINGS

Once transplanted, your seedlings are at risk not only from unseasonal changes in temperature, but also from marauding pests, such as slugs and snails. Have an old sheet to hand so that you can quickly cover seedlings if the temperature is forecast to drop unexpectedly (see my four-degree rule, p24). It's also a good idea to set slug traps and barriers to prevent your seedlings from being munched overnight (see pp160–163).

4

5

HOW TO WEED

Weeds may be nature's way of protecting bare soil, but these tenacious plants can quickly take root in your carefully cultivated ground. Below are my tried-and-tested ways of keeping your growing space weed-free.

Weeds can establish very quickly so the best strategy is to pull them out as soon as they appear. The longer you leave them, the stronger their root systems get and the harder they are to remove. It's also vital to pull up weeds before they can set seed and multiply throughout your garden. I set aside time each week to have a quick scout around and pull out any new weeds that have appeared. It's also important to avoid digging, raking, or disturbing the ground unnecessarily. This will simply bring viable weed seeds to the surface of the soil, where they will germinate and grow.

OUTCOMPETE THEM
One of the most effective ways of reducing weed numbers is to keep the ground densely planted so they struggle to grow. Once your vegetables are well established they will cast a lot of shade. These are not ideal conditions for weeds and will result in reduced germination and weaker seedlings.

Planting methods such as intercropping (see p20) are also a great way of minimizing gaps between plants, while at the same time maximizing the productivity of your patch. I like to plant radishes among onions, for example, to get two different harvests from the same space.

Succession sowing (see p42) is another good option because you are making full use of your growing space. Having bare soil for long stretches of time will certainly mean that weeds are more likely to grow.

COVER THE SOIL
Mulching the soil can prevent weeds colonizing bare patches of ground. Any that do make it through the mulch are easily removed. A home-made compost mulch will also add nutrients. Ground that isn't in use can be covered with a few layers of cardboard or newspaper, with stones on top to keep the material in place. One drawback of this method is that you unwittingly create a slug habitat, but there is a solution. Make a cardboard slug trap (see p162) and you will be controlling both slugs and weeds. Weeds still grow during winter, so it's always a good idea to cover any bare soil at this time of year.

DISPOSING OF WEEDS
Leaving uprooted weeds to bake in hot sun will certainly kill them. During cooler weather, collect weeds that haven't yet flowered and put them in your compost bin. Don't ever add weeds that are in flower or have gone to seed. If weed seeds are in your compost when you come to apply it to the ground, you will inadvertently be spreading them across your growing space.

TIP
Weed on a sunny morning then leave the small, uprooted plants on the surface to dry out and die. The material will break down and return nutrients to the soil.

Pull out any weeds as soon as they appear (*top left*). Strategies to prevent weed growth include planting densely to crowd them out (*top right*), or suppressing them with cardboard (*below left*).

EDIBLE WEEDS

Dandelions (*see above*), chickweed, and nettles are good sources of free food, but will soon take over your bed if left to grow. Instead, if you have the space, set aside a corner for wild edibles that need no input from you to grow and where you can forage. Nettles are also valuable because they attract beneficial insects to pollinate plants, and birds to prey on pests.

HARVESTING AND STORING

Eating food you have grown yourself is like having access to your own free greengrocer 365 days a year. The sense of comfort, and not to mention the personal pride, in having something ready to harvest whatever the season is also hugely motivating.

If you are just beginning the food-growing journey, you will need to be patient. Waiting for plants to grow and mature can be a challenge, but as soon as you harvest and eat your first crops, you'll know it was worth it. Most gardeners experience gluts of produce, so why not share something from your harvests with a friend or a neighbour – you might also inspire them to grow their own food.

HARVESTING YEAR

Spring can be light on harvests. Some crops left over winter will reappear in early spring, like chard, and the first of those planted this year will come through from mid-spring – look out for radish, lettuce, and newly planted spinach.

Summer sees the start of the major harvests. Most of your crops will be ready during this period: peas and beans, root veg, summer squash, salad leaves, tomatoes and peppers, as well as garlic and onions.

Major harvests continue until late autumn, when winter vegetables, such as swede and leeks, start to make an appearance.

Make sure you harvest pumpkins and the rest of your winter squash in time for Halloween! Wait until after a frost before your pull out your parsnips – one of my favourite vegetables!

By midwinter there's not a great deal left to harvest, but you can continue to take swede, leeks, and kale leaves as you need them.

PLANNING FOR THE HUNGRY GAP

The period between early spring and early summer when winter vegetables have finished but new crops haven't yet matured is known as the "hungry gap". Make sure to save some of your winter veg for this period. If you leave the roots of Swiss chard and perpetual spinach in the ground over winter, plants will send up a new flush of delicious fresh leaves when the weather warms up in spring.

STORAGE

Eating food as soon as possible after harvesting is the best way to get the most flavour from your crops. Most can be stored in an airtight container in the fridge for five days to a week after harvest, but check the options below for other ways to store crops, and the types of annuals most suited to this form of storage.

- **Leaving crops in place** is perhaps the easiest means of preserving flavour. You can use this method for hardy vegetables, such as kale, parsnips, and swede. Carrots and beetroot can also be kept in the ground, but should be mulched if still in place by early winter.
- **Blanch and freeze** harvests of Brussels sprouts, peas, and Swiss chard stems that you can't eat at the point of harvest. To blanch, cook vegetables in boiling water for a couple of minutes, and then immediately transfer to cold water. Drain, leave to dry, and then freeze them.
- **Dry potatoes, chickpeas, onions**, and garlic before storing them. Details are given for each crop in its specific growing instructions.

Nothing beats the feeling of returning from your veg patch with a bumper harvest of fruit and vegetables.

TIP

Failures can occur, however much growing experience you have, but you will always get another chance the next year. Learn from your failures, and enjoy your successes.

Dwarf French beans

TIP

*Avoid saving seeds
from crops that have bolted
(flowered and set seed early).
These plants may produce
offspring that are similarly
prone to bolting.*

Pinto beans

Garlic cloves

Chilli seeds

Nasturstium seeds

Peas

Beetroot seeds

Lettuce seeds

Radish seeds

SAVE YOUR OWN SEED

When it comes to growing food for free, there are two key reasons for saving your own seed. First, seed-saving reduces the need to buy in supplies; second, you can barter it for other varieties at seed swaps – at no cost.

Seed, in a general sense, refers to those parts of the plant that produce offspring, and I include cloves (garlic) and tubers (seed potatoes) for simplicity. Some annuals produce flowers and set seed in the year that they were sown. For others, like beetroot, you have to wait until the year after before crops set seed for you to collect. I save seed from:

- all beans and peas
- radish
- potatoes
- oca
- lettuce
- spinach
- tomatoes
- peppers
- garlic

Seed-saving instructions for each of these crops can be found in that plant's specific growing instructions on the following pages. When selecting plants to save seed from, make sure you pick your best specimens. Seed from these is most likely to result in high-quality crops.

TRUE TO TYPE

Many vegetables, such as carrots and kale, readily cross-pollinate with other plants around your garden and community. This means that you can't be sure of the quality of the crops their seed will produce, as they will contain genetic material from the plants with which they have cross-pollinated. For this reason, I haven't provided seed-saving instructions for these crops.

Some vegetables, such as tomatoes, beans, and peas are self-pollinating, so their seed will almost always produce the same variety of plant as the one it was collected from. There is a small chance of cross-pollination, but this is a risk I will always be willing to take.

Hybridized vegetable varieties (such as seeds labelled F1) result when breeders deliberately cross-pollinate different plants. Seed collected from these varieties will not give you consistent results, so always stick with named varieties if you want to save seed after harvesting. These will be labelled "heritage", "heirloom", or "open-pollinated" on seed packets.

STORING SAVED SEEDS

You must dry seeds out for a few days to remove as much moisture as possible – ideally leave them on newspaper or a baking tray on a sunny windowsill. Then place them into old envelopes, or pots and glass jars with airtight lids. Don't forget to label the containers because some seeds look very similar and are easily mixed up. Seeds in sealed transparent containers are best stored in drawers and dark cupboards. Seeds in envelopes or containers that exclude light can be kept in the fridge. Make sure you write the date you saved the seeds on each envelope, and check expected longevity for that crop so that you know how long the seeds will remain viable.

When the plants grown from your saved seed begin to lose vigour and are producing unusually low yields, it's time to sow fresh seed. You can source these for free from seed swaps.

I save seed from a variety of plants, including beans, garlic, peas, tomatoes, lettuce, and radish.

PEAS AND BEANS

All podded vegetables are happy in the ground, raised beds, or large containers. The taller-growing legumes, such as peas and climbing beans, maximize vertical space – a massive benefit if you only have a small area.

Beans and peas tend to suffer less from pests and diseases than other crops, especially if you start them in modules indoors, where they are also protected from the cold (some are frost-tender). At the end of the season, when you've harvested your entire crop, cut the plants down but leave their nitrogen-fixing roots to enrich the soil.

a. RUNNER BEANS (AND OTHER CLIMBING BEANS)

Originally grown as an ornamental for their red flowers, runner beans produce delicious, abundant crops and need very little maintenance.

Start off

In mid- to late spring, sow seeds indoors in cardboard tubes, so the long roots have space to grow, and won't get damaged when the seedlings are transplanted. Use compost or a 50/50 compost-soil mix, and push one seed per tube down to a depth of 4–5cm (1½–2in). Place on a sunny, indoor windowsill and keep the compost moist.

Grow on

Runner beans are frost-tender, so wait until the risk of frost has passed and plants are over 10cm (4in) tall before transplanting the whole tube into your outside space. Growing 2m (6½ft) tall, climbing beans need a supporting structure (*see right*). Give each plant its own cane or stick, tying it gently to the support with string to encourage it to twine. Water well when flowers form and cut off the growing tip (the top) of each plant when it reaches the height of the support.

Harvest

From late summer onwards, pick your beans when they are about 20cm (8in) long and still tender – about three months after sowing. As long as there are still flowers, you'll get more beans. When beans get larger and slightly leathery, de-string them by trimming off the tough edges before cooking.

Save seed

Leave a few pods on the plant to turn brown, then remove them from the plant and take out the seeds. Dry the seeds and store them in an envelope (see p129). Use within three years.

CREATE A SUPPORT STRUCTURE FOR BEANS

Make a simple wigwam by pushing tall (about 2m/6½ft), stout sticks or bamboo canes 30cm (1ft) into the ground and tying them securely together at the top. Otherwise, build an A-frame (*see below*) with two "ends" and vertical supports every 30cm (12in) joined by a horizontal "ridge".

b. DWARF FRENCH BEANS

Growing only to 45cm (18in) tall, French beans don't need supports, but do need protection from cold.

Start off

In May, sow seeds in pots or cardboard tubes as for climbing beans and place on a sunny windowsill. Keep the compost moist.

Grow on

After the last frost and when seedlings are about 10cm (4in) tall, remove them from their pots and transplant them into your outside space (or plant the whole lot if the container is biodegradable). Space them fairly close – 15cm (6in) apart – so they support each other. If any are getting weighed down by pods, push in a stick or twig 5cm (2in) from the plant to provide support. As they are small plants with shallow roots, I mulch my dwarf French beans with a 2–3cm (1in) layer of grass clippings during the growing season. This retains soil moisture and prevents the roots drying out.

Harvest

From late summer, Start picking once the beans are 8–10cm (3–4in) long, generally two to three months after sowing, and before their skins become tough. Steam or lightly boil before eating to preserve the goodness.

Save seed

As for runner beans (*see opposite*).

c. BROAD BEANS

Broad beans, unlike climbing beans and dwarf French beans, are hardy and easy to grow. Some varieties have pretty red flowers.

Start off

In late winter or early spring, sow one seed per compost-filled cardboard tube (these will allow enough space for the long roots to grow). Place on a sunny windowsill and water regularly.

Grow on

Transplant the seedlings in their tubes when they are 7cm (2¾in) tall, allowing a 30cm (12in) gap between

a

plants. Broad beans have long roots that may poke out of the tube. Try to avoid damaging the roots, but the plant will still grow if they break. In exposed locations or if strong wind and rain are forecast, support the stems by inserting a 60cm (2ft) cane or stick about 5cm (2in) from the plant and tie loosely with soft string or strips torn from old sheets.

Harvest
Remove the tops of the beans (top six or so young leaves) once they have started flowering to prevent blackfly infestation. Eat the tops lightly cooked or stir-fried. Pick the beans when the pods are firm and the beans are a good size. Crops mature from seed within three to four months.

Save seed
As for runner beans (see p130).

a. PEAS
My peas hardly ever make it into the kitchen because they are the perfect sweet garden snack. Some peas can grow more than 160cm (5ft) tall and the tendrils do need supports to latch on to. Peas will also survive light frosts and can be sown in early spring.

Start off
Sow peas in cardboard tubes or small pots – their roots don't need as much space as broad beans – from

PEA STRUCTURE
Insert twiggy sticks (one per clump) to support seedlings for the first 30cm (12in) of growth. I use old Christmas-tree branches and two recycled fence posts with wire in-between as the main structure (*see right*). If you are growing peas in containers, push in branches or surround the pot with wire mesh. Different pea varieties grow at different heights so if you are unsure, aim high. Once the peas have reached their full height, make a note of it. Next year, when you sow seed you have saved from your crop, you will know how tall to make the supports.

early spring to midsummer. Fill with compost; sow two to three seeds per module and don't bother thinning – the plants won't compete with each other as they mature. Keep moist and place on a sunny windowsill. You can also sow peas to harvest the shoots. Fill a shallow tray with drainage holes to 5cm (2in) with compost, then add a layer of peas, followed by a 1cm (½in) layer of compost. Water well and leave on a sunny windowsill.

Grow on
If growing for peas, transplant seedlings when they are 7–10cm (2¾–4in) tall. The clumps can be spaced closely at just 5cm (2in) apart. You will need to add a support structure for them to grow up (see *opposite*).

Harvest
You can harvest pea shoots when they reach 6cm (2½in) tall – add them to salads raw. Even if you are growing peas to maturity, you may want to pick out some of the shoots to encourage bushy plants. Pick mature peas when their pods are firm to the touch. This should be 12–14 weeks after sowing.

Save seed
As for runner beans (see p130).

b. CHICKPEAS
Chickpeas will withstand only light frosts but, unlike other legumes, they cope well in droughts. Chickpeas flourish in the UK climate, especially when the summer is warm and sunny.

Start off
Sow in cardboard tubes or small pots indoors in mid-spring for the best results. Insert one seed per pot or tube, and keep them on a sunny windowsill.

Grow on
Plant out seedlings when they are at least 7cm (2¾in) tall. Space plants 15cm (6in) apart so they support each other as they grow. Chickpeas reach a height of just 45cm (18in), so don't need a structure for support.

Harvest
Chickpeas take about three months to mature from seed. Squeeze the green pods, and if they feel firm at touch, then they are ready to be harvested.

Save seed
As for runner beans (see p130).

b

Each chickpea pod contains two peas on average, so you'll need to grow a reasonable number of plants to get a good harvest.

BRASSICAS

Vegetables in this group are all sown at roughly the same time using the same method. The lush green growth needs rich soil, so make sure you incorporate plenty of compost or well-rotted manure before you plant out seedlings.

Start off these hardy (cold-tolerant) plants in modules or pots to be transplanted later. If you have space, sow the seeds directly into the ground, then thin out. Brassicas can be targeted by cabbage root fly and cabbage white butterflies. See p164 for controls. I've left out cauliflowers as these take up a lot of space and can be tricky to grow, so maybe try these once you've built up some confidence.

a. BRUSSELS SPROUTS
This vegetable certainly divides opinion but home-grown Brussels sprouts really are hard to beat, especially picked fresh on Christmas morning.

Start off
In late winter, sow three seeds per small pot or module filled with a 50/50 mix of soil and compost, and place on a sunny windowsill. Thin to the strongest seedling per pot a week after they appear. Alternatively, thinly sow a row of seeds 2cm (¾in) deep directly into a raised bed or container. There's no need to thin, just lift the seedlings when they are 7–10cm (2¾–4in) tall, and transplant them into their final position.

Grow on
Transplant seedlings, spacing them around 30cm (12in) apart and water regularly. If the plants look top heavy and unstable, stake them in autumn.

Harvest
There are early, mid-season, and late-cropping varieties of Brussels sprouts. Harvest them when they are a decent size and feel nice and firm. You can harvest the leaves to make the crop go a little further – steam or stir-fry them before you eat them.

b. CABBAGE
I find summer- and autumn-cropping cabbages give me the best yields, but you can also grow winter cabbages, which tend to have smaller heads so are less likely to be blown over by wind.

Start off
As for Brussels sprouts. Sow summer cabbages in late winter, but wait until spring to sow autumn cabbages.

Grow on
Transplant seedlings so that they are least 45cm (18in) apart to give them space to spread out. Because the roots are small in comparison to the top growth, some cabbages can get blown over during storms. Carefully firm in the soil around the seedlings once they have been transplanted and water regularly so that they don't dry out.

Harvest
Pick cabbages when the heads are large and feel firm when you squeeze them.

SAVING BRASSICA SEED

It's not practical to save brassica seed, as cross-pollination and inbreeding can lead to poor-quality cropping. Try to get brassica seeds at a seed swap – one packet can keep a small garden going for a few seasons, and the seeds will stay viable for up to four years.

a

b

c

d

c. KALE

To me, kale is the king of the vegetable garden: it is extremely productive, very hardy, and both the flowers and leaves are edible. Kale grows well in large containers if you don't have much space.

Start off

As for Brussels sprouts. Sow between early spring and early summer.

Grow on

Transplant kale seedlings when they are 7–10cm (2¾–4in) tall, spacing them 45cm (18in) apart to give each plant plenty of room to grow. In exposed or windy sites, plant slightly closer together. Kale copes with drought fairly well.

Harvest

From mid-autumn onwards, start harvesting the lower leaves, picking little and often. Taking too many at once will affect the growth. Leave in the ground over winter, and let the plant flower in spring. Treat the flowers in exactly the same way as purple sprouting broccoli, picking the shoots as they appear. You can expect to harvest these shoots for weeks.

d. PURPLE SPROUTING BROCCOLI

For many, this is their number-one vegetable. I grow the late-cropping variety because it can be harvested in winter and early spring, when there aren't many other fresh vegetables around.

Start off

As for Brussels sprouts. Sow seeds in spring.

Grow on

Plant out seedlings, spacing them around 45cm (18in) apart so they support each other during winter when mature. Water regularly, especially during dry spells.

Harvest

Cut off the purple shoots as they appear. I find the more you harvest, the more shoots the plant sends up. Harvest the leaves too – they taste good steamed or chopped up in stir-fries.

ROOT VEGETABLES

As the name suggests, these veg grow underground and some can stay there even when there is frost on the surface. Some of those mentioned are technically brassicas, but I treat them as root vegetables.

a. RADISH

These small roots couldn't be easier to grow and are excellent for containers.

Start off

Radish seed can be sown from early spring to late summer, and even, at a push, into early autumn in warmer regions. Sow seed thinly in a container and cover with 1cm (½in) of compost. In beds or in the ground, sow thinly in rows 1cm (½in) deep and 10cm (4in) apart.

Grow on

I don't thin radish because the plants create space as they grow by pushing away from one another. Sow one row every week or two to avoid a glut.

Harvest

Radish mature within four to six weeks, and are best eaten very fresh. Check the size and gently pull up the largest by the base of the leaves, leaving the smaller ones to grow on. Leave some to form seed pods, which can be eaten when they're green, or leave them to dry and collect the seed (*see below*). You can eat the leaves when roots stop forming in late autumn (or earlier). Pull up your crops once frosts arrive.

a

SAVE RADISH SEED

Leave six radish plants you sowed in early spring in the ground to flower and set seed. When the pods turn brown (*see below*), uproot the plants and hang them up in bunches, pods pointing downwards. Place a box beneath to catch the seed when the pods dry out and open. Store the seed in sealed glass jars or envelopes in a cool dark place, and use within two years.

b   c

b. BEETROOT

Beetroot are one of my favourite vegetables to grow and eat. The roots are delicious roasted and the fresh young leaves can be eaten raw.

Start off

Soak the seeds in water before sowing for quicker germination. Sow outdoors from early spring right through to midsummer in rows 2cm (1in) deep and 20cm (8in) apart. You can also start off beetroot in newspaper pots filled with compost (see p30). Beetroot is hardy and the seeds remain viable for three years.

Grow on

Thin to one seedling every 5cm (2in) or transplant seedlings into their final position when they have three or four leaves. Water regularly, especially in dry weather.

Harvest

Pick a few young leaves and stir-fry or add to salads. I usually pull up beetroot before it gets to the size of a tennis ball and use by midwinter. Spring sowings will be ready in summer, and summer sowings will ready in autumn.

c. CARROTS

The thrill of pulling up carrots is hard to describe, but believe me, it's one of the highlights of my year.

Start off

Carrots can be grown in the ground, raised beds, or large containers but struggle in smaller ones. They do not transplant well, so sow them directly into their final position, as thinly as you can, from mid-spring to midsummer in rows 1–2cm (½–1in) deep and 15cm (6in) apart. Don't allow the soil to dry out during the germination process. Seeds stay viable for three years.

Grow on

Thin to one seedling every 5cm (2in) and take steps to prevent carrot fly, a common pest that can wreak havoc (see p164). Water well in dry spells.

Harvest

The diameter of the orange top of the carrot sitting just above the soil is a good indication of the size of the root below. When you're happy with their size, you can pull up the largest roots first – leaving the smaller ones in place to continue to grow. Although hardy, I tend to cover carrots with a layer of leaves or straw to protect them and lift when needed.

a. TURNIPS

Even if you think you don't like turnips, try growing your own and eating them young. You may be pleasantly surprised.

Start off
Sow seeds directly into the ground or a raised bed, one seed every 1cm (½in) and in rows 1cm (½in) deep and 15cm (6in). Thin to one plant every 5cm (2in). You can sow early varieties from early spring, and late varieties from midsummer. The seeds will remain viable for four years.

Grow on
Water regularly, especially in dry weather, and keep on top of weeding.

Harvest
Turnips mature after just 8–10 weeks and are best enjoyed young before the flavour gets too overpowering. I lift all turnips by midwinter. They'll keep in the fridge for up to a week.

b. SWEDE

Swede has just half the calories of potatoes so makes a fantastic alternative. The root sends up new edible leaves in spring if left in the ground over winter.

Start off
Sow seeds in late spring directly into the ground or a raised bed at 2cm (1in) apart in 2cm (¾in) deep rows 30cm (12in) apart. Seeds stay viable for four years.

Grow on
Thin to one seedling every 7–10cm (2¾–4in). Weed and water regularly.

Harvest
From mid-autumn onwards, lift swedes, largest first, as you need them or leave in the ground over winter.

c. PARSNIPS

Although parsnips are slow to germinate and develop, they are delicious roasted and always taste sweeter after the first frost.

Start off
From early to mid-spring, sow three seeds every 7–10cm (3–4in) directly into the ground. Rows should be spaced 20cm (8in) apart. Don't allow the soil to dry out while the seeds are germinating.

Grow on
Thin to the strongest seedling and weed regularly. When growing well, water only if the soil is very dry.

a   b

Harvest

Pull up parsnips, largest first, from autumn until late winter. Wait until after the first frost for the best-tasting roots.

d. POTATOES

Potatoes are very easy to source for free. In one year I received three batches of the sprouted tubers from the back of friends' cupboards. There are three types: first earlies (aka "new potatoes"), second earlies, and maincrop. Potatoes grow best in a location that enjoys full sun for most of the day and do well in large containers, raised beds, as well as in the ground.

There are three methods I use to grow potatoes – dig, no-dig, and the container method. These are all explained below.

Start off

Begin the growing process for all three methods with seed potatoes. "Chitting" the potatoes first, which means encouraging shoots to form, will give your crop a head start but isn't essential. Two to three weeks before planting, put the seed potatoes in a single layer (cardboard egg boxes are perfect) in a light, cool place so the end with the most "eyes" is facing upwards. They will be ready to plant out when they have short, strong shoots about 3cm (1in) long.

Transplant

First earlies are planted in early spring (ready to harvest in 10–12 weeks), second earlies in early to mid-spring (ready to harvest in around 14 weeks), and maincrop potatoes in early to mid-spring (ready to harvest in 16–18 weeks).

Potatoes can be planted later but this increases the risk of potato blight (see p166). In wetter regions, I recommend planting no later than mid-spring, prioritizing first and second earlies.

- **Dig method** You will need raised beds at least 30cm (1ft) deep. Dig a trench about the depth of a spade (25cm/10in) and add a 5cm (2in) layer of well-rotted manure, coarse compost, or vegetable scraps to the base. Place one seed potato every 30cm (12in) along the trench then cover the potatoes with the soil from the trench. If you have space for more rows, leave 30cm (12in) between them. Shoots should appear in three to four weeks.
- **No-dig method** You can grow potatoes using this method either in prepared ground (see p40) or in a shallow raised bed with topsoil to a depth of at least 10cm (4in). Make sure grass can't creep into the bed. Place seed potatoes 30cm (12in) apart on top of the soil and cover with at least 5cm (2in) of compost or well-rotted manure. Then cover the compost with

c

d

a thick layer (20cm/8in) of shredded leaves, straw, or a mix of grass clippings and autumn leaves.

- **Container method** Large containers (around 30 litres/6½ gallons) are perfect for growing potatoes. Half fill with a 50/50 mix of topsoil and compost/well-rotted manure and plant two to three seed potatoes on top. Add more of the soil/compost mix, leaving a gap of 7cm (2¾in) below the rim of the container.

Grow on

Potatoes are frost-tender, but a light frost will kill only the growth above the ground, and the plant will send out new shoots. See pp24–25 for ways to protect your crop from frost.

- **Dig method** When the shoots have grown to around 7cm (2¾ in) above the surface, use a rake to mound soil over the top of them This is a process known as earthing up and will help create greater yields. You only need to do this once. Water your plants if the soil is dry.
- **No-dig method** Once the potato shoots emerge, cover with another 15–20cm (6–8in) layer of mulch.

Potato flowers appearing is a **sign** that your crop is ready to harvest.

- **Container method** When shoots appear, fill the container to the top to cover them, using the same mix as when you sowed your potatoes. Keep well watered when the plant is growing strongly, and especially in warm weather.

Harvest

Once your potatoes start to flower, gently remove some compost from the roots to see if there are any tubers worth harvesting.

- **Dig method** Dig up one plant with a fork. If you think the potatoes are too small, leave the remaining plants for another week.
- **No-dig method** Simply pull away the mulch to reveal the potatoes. This method of growing potatoes means you can take only a few tubers from each plant at a time.
- **Container method** Gently tip the container to release the contents into a trug or onto an old sheet and pick out the potatoes. Add the remaining soil and compost to a raised bed, or mix with some extra compost and use in another container.

Storage

Don't wash your harvested potatoes. Instead, leave them outside on some cardboard or newspaper to dry in the sun for a day. Put them straight into hessian bags or cardboard boxes lined with newspaper and store in a cool, dry, dark place.

SAVE SEED POTATOES

When you harvest your crop, select some small to medium-sized potatoes to use as seed potatoes for planting next year. They should be taken only from plants that haven't been exposed to blight. Dry them outdoors in the sun for two to three days then put them in a cool, dark, airy place such as a shed or garage. Store them in cardboard egg boxes with a couple of sheets of newspaper laid over the top to encourage good airflow around the tubers.

DID YOU KNOW

Potatoes are perennial plants, which is why any tubers left in the ground may sprout the following year. They can suffer from a range of pests and diseases so are best treated as annuals.

Oca, are small tubers that are often pink or reddish in colour (*above*). Once the bushy plants appear (*left*), they can be earthed up in the same way as potatoes. Look out for the yellow flowers (*above left*) in summer.

OCA

Like potatoes, oca is also a perennial that I treat as an annual. Although originating in Central America, oca was introduced to New Zealand, which explains why it's known as the New Zealand yam. Growers get very excited about this relatively uncommon vegetable and will often give away a few tubers if you show some enthusiasm. Oca tubers are small and have a lower yield than potatoes. Like potatoes, they are grown each year from tubers that you plant in spring. Oca plants are ideal for smaller spaces because they are relatively short and grow well in large containers and tyre planters. They do, however, need a spot that gets plenty of sunlight.

Start off

Plant oca tubers outside after all risk of frost has passed. Use a stick to make holes around 5cm (2in) deep and 20cm (8in) apart in containers or raised beds filled with a mix of at least 70/30 topsoil and compost (although the more compost, the better). Drop an oca tuber down into each hole and use your hands to cover over with soil. You can also plant tubers 5cm (2in) deep in pots on your windowsill in early spring and then transplant the seedlings outside, 20cm (8in) apart around two to three weeks after your average last frost date. If an unexpected frost is forecast, you will need to take action to protect the tender oca shoots (see pp24–25).

Grow on

I recommend earthing up oca, in the same way as potatoes, by gently covering the plants with a 5–10cm (2–4in) layer of soil when they are around 7cm (3in) tall. The tubers grow near the surface so this gives them more space and can help to increase the yield. The only task remaining is to keep weeds down in the early stages of growth and to water if the soil becomes very dry. By midsummer, the plants have bulked out and spread, which effectively keeps weeds at bay.

Harvest

After the first frost, the leaves and shoots will start to die back, but the tubers will stay protected in the ground. Wait until all the top growth has died and started decomposing (usually late autumn to early winter) before harvesting all the tubers – leave it any later and mice may decimate your harvest.

Saving seed

Like potatoes, you need to store the oca tubers you want to plant next year in a cardboard box lined with a couple of sheets of newspaper over winter. Keep the box in a cool (but frost-free), dark place until you are ready to plant them the following spring.

CHINESE ARTICHOKES

These tubers (*right*), like oca, are fairly difficult to source for free. Interest is building however, and I wouldn't be surprised if a large number of gardeners start growing them over the next few years. Chinese artichokes can be grown in exactly the same way as oca. It's also important to harvest all of your Chinese artichoke tubers in one batch to avoid the disappointment of rodents eating the tubers over winter. Save some harvested tubers for growing next spring, in the same way as oca (*see above*).

SALADS AND LEAF CROPS

Salad leaves are expensive to buy considering how much can be grown from a single seed packet. Treat them, and the other leafy veg below, as cut-and-come-again crops by taking a few leaves at a time. Fresh leaves will soon appear.

You can grow all of the salad and leaf crops mentioned here in containers. Fill a container with compost, sprinkle the seeds over the surface, and cover with another layer of compost (about 1cm/½in). Keep well watered, and pick the largest leaves as you need them, allowing the smaller leaves to develop.

a. LETTUCE
Harvest leaves of densely sown lettuces as you need them. They can be grown in the ground, raised beds, or containers.

Start off
Sow lettuce seed directly into their position outdoors from early spring through to late summer in rows 1cm (½in) deep and 10cm (4in) apart. Seeds will remain viable for up to five years.

Grow on
Water well, especially before seedlings show, taking measures to deter slugs (see pp160–163). To reduce bolting, grow crops in partial shade.

Harvest
Cut leaves as you need them and before the crops bolt, when the leaves turn unbearably bitter.

b. SPINACH
In warmer regions, try a late sowing of spinach in raised beds and keep them in place over winter so that they crop again in spring.

Start off
Sow directly into their position outdoors from early spring to late summer in rows 2cm (¾in) deep and

a

b

c

15cm (6in) apart, or sow individual seed in modules. Seeds remain viable for three years.

Grow on
Transplant seedlings 10cm (4in) apart once they have four to five true leaves. Spinach sown in late summer can be transplanted into raised beds overwinter (unless your garden regularly experiences frosts).

Harvest
Pick young leaves for salads as soon as they are large enough or wait 10–12 weeks for bigger leaves. Spinach will bolt but the small leaves are still edible. Overwintered spinach can be picked from early spring.

c. ROCKET
Fiery rocket leaves add a kick to salads and you can sow them up until late summer.

Start off
Mid- to late summer sowings produce the best crops, but you can start in mid-spring. Sow in rows 1cm (½in) deep and 10cm (4in) apart in the ground or a raised bed.

Grow on
Keep rocket well watered in dry spells. It has a tendency to bolt.

Harvest
Pick young leaves as and when you need them for a cut-and-come-again crop. The summer flowers are also edible, with a strong peppery flavour.

d. SWISS CHARD (AND PERPETUAL SPINACH)
My favourite kind of chard is 'Bright Lights', a mix with amazing yellow, pink, and red stems.

Start off
For baby leaves, sow seeds direct in rows 10cm (4in) apart. For larger leaves, sow three seeds 2cm (¾in) deep every 15cm (6in). The seeds will remain viable for up to three years.

Grow on
Keep chard well weeded and watered in dry spells. If you sowed seeds in threes, retain only the strongest seedling – it will produce the best-quality plant.

Harvest
Cut tender young leaves regularly to add to salads. Plants take around ten weeks to mature, by which point both the leaves and stems are delicious lightly cooked; continue to harvest for months. When finished, mulch the crowns with autumn leaves and leave the plants in place over winter. The plant will send up a flush of new leaves in spring before bolting.

d

SAVING SEED
Leave a few lettuce and spinach plants in the ground to flower. Once the flowers turn brown, dig up the plants and hang them upside down with a pot or box beneath to catch the seeds when they fall. Alternatively, once the pods have dried, run your fingers along the stems and catch the seeds as they fall (*see right*).

CUCURBITS

This productive group of vegetables, known collectively as cucurbits, can fill space very quickly. Plant a few to get excellent yields and lots of weed-suppressing foliage, which will also help prevent the soil drying out.

All cucurbits need full sun and are frost-tender. Wait until after the last frost before transplanting them outside and give winter squashes enough time to ripen before mid-autumn – even a light frost will kill them. Cucurbits like rich soil and are thirsty plants, so never let them dry out. Some squashes are bushy, but others trail and can soon take over small spaces. Courgettes and cucumbers grow well in large containers.

a. WINTER SQUASH AND PUMPKINS

The climbing stems of some pumpkins, can reach up to 5m (15ft). Circle them around short sticks to save space.

Start off

In early spring, germinate seeds using the germination test (see p83). When the first tiny leaves appear, gently tear the paper around the roots of the germinated seed, and plant the seed and section of paper in an almost-full pot of compost – this should be around midspring. Cover the germinated seed with around 1cm (½in) of compost, then water well, and leave on a sunny windowsill.

Grow on

Transplant outdoors when plants have at least four true leaves, but only once there is no risk of frost – early summer is an ideal time for this. Squashes are hungry plants, so add two to three large handfuls of compost or well-rotted manure at the base of the planting hole. They grow slowly for a few weeks after being transplanted.

a

Harvest

When ripe and fully coloured, cut the stem but not too close to the fruit. Harvest unripe pumpkins before the first frost and leave on a sunny windowsill to mature.

Saving seed

Although their seeds stay viable for five years, squashes cross-pollinate, so saved seed may not grow true to type.

a. COURGETTES

Courgettes are the easiest and most productive of the cucurbits. Some are yellow or striped; others are round.

Start off

Sow seed as for winter squash (*see opposite*), or in compost-filled pots indoors at a depth of 2cm (1in) in mid-spring. Seeds are viable for five years.

Grow on

In early summer, when courgette seedlings are about 15cm (6in) tall, transplant them outdoors. Protect against damage from slugs by placing bramble twigs around the base of the plants, and check for slugs at night. To grow in containers, give each plant a large (45cm/18in across), compost-filled pot. Keep well watered; avoid splashing the leaves to prevent powdery mildew (see p167).

Harvest

From midsummer, pick courgettes regularly to extend the harvest to early autumn.

b

c

a. CUCUMBERS

Cucumbers can be grown under cover, but I always recommend you choose a well-known outdoor variety.

Start off

Sow cucumber seeds like winter squash seeds, or indoors in compost-filled pots in early spring to increase chances of a successful harvest. Seeds are viable for five years.

Grow on

Transplant into large (45cm/18in across), compost-filled containers and keep on a sunny windowsill. In early summer, place containers outside, ideally against a south-facing wall, and provide support. Train the stems up a small wigwam (around 1m/3ft tall) made from sticks or canes with string wrapped around and keep well watered.

Harvest

Pick regularly from midsummer to encourage more fruits.

TOMATOES

There are two types of tomato plant, bush and cordon (or vine). Bush tomatoes don't need supports, but take up more space than cordon. Most of the popular tomato varieties are cordon types, and I recommend you start with these.

Tomatoes relish a warm, sheltered, sunny location. Sowing seed indoors in late winter or early spring will give the plants a fantastic head start and increase yields over a longer period. Tomatoes are frost-tender, so don't transplant them outside until all risk of frost has passed. A light-coloured, south-facing wall is the ideal spot to grow tomatoes, and in medium to large containers (a tyre planter will also work well) filled with a 50/50 mix of topsoil and (preferably) homemade compost. You can source seed by saving them from ripe, shop-bought tomatoes or from a friend's home-grown fruit (*see opposite*).

Start off

Old yogurt pots with holes at the base are ideal for sowing tomato seeds. Add a 2.5cm (1in) layer of compost and set three seeds equally spaced on top. Cover with 5mm (¼in) of compost and don't let it dry out. You'll get between one and three seedlings. Remove all but the strongest and let this grow on to about 5cm (2in); it will look a bit leggy. Gently fill compost around the seedling up to the first set of leaves to create a strong plant. Seeds are viable for four years, but some can last over a decade.

Grow on

If the tomato seedling is outgrowing its pot, gently remove it from it's pot – turn the container upside down and place your fingers over the top. Tap the base firmly to release the roots. Then, transplant it into a larger pot filled with compost to develop further. When the weather starts warming up and you haven't had a frost for at least three weeks, it's time to transplant the tomatoes into their final growing position. Fill your chosen pot with the 50/50 mix of soil and compost and insert one tomato plant per container.

Cordon tomatoes need a strong vertical support at least 1.5m (5ft) in height. A bamboo cane works well or you could lean a pallet up against the wall. As the stem grows, use soft string to tie the plant to the support at intervals of 10cm (4in). Cordon tomatoes develop sideshoots – pinch them out with your fingers. This will focus the plants energy on producing flowers and not more foliage. The removed sideshoots will take root if you pot them up, but the plants won't develop fruit before temperatures drop. When four sets of flowers (trusses) are forming small fruits, cut off the top of the stem to stop further growth and give the fruit the best chance of ripening.

Irregular watering can cause tomatoes to crack and give rise to a disorder called blossom end rot, which affects the base of the fruit (see p166). Prevent problems by watering well every two to three dry days, and after heavy rain wait for two or three days before watering again. Tomatoes are hungry plants and need feeding once every week or two after the first flowers have appeared. Homemade comfrey feed (see pp74–75) is ideal and rich in nutrients. Dilute it in a watering can.

Harvest

Pick tomatoes when they turn red (or yellow or orange, depending on the variety). Store them on the windowsill and eat within a few days. I avoid storing them in the fridge as I find this reduces their flavour – although you may need to if you have too many to eat before they start to turn.

I grow a wide range of tomatoes and particularly look out for seeds of those that you can't find in supermarkets.

SAVE TOMATO SEED

Pick ripe tomatoes and scoop out the seeds into a clean glass jar. Half-fill it with water and leave until a scum forms on the surface (around five days). Remove the scum then empty the contents into a sieve. Rinse under the tap, separate out the seeds, spread them on a paper towel, and leave on a windowsill. Once the seeds are thoroughly dry, collect and store in a sealed glass jar or envelope in a cool, dark place. Use within four years.

PEPPERS AND CHILLIES

These colourful members of the capsicum family are perfect for containers.
They don't take up much space and can be grown indoors on a sunny windowsill.
You can even start them from seed saved from shop-bought fruit.

Peppers require heat and humidity. In cooler regions, grow them inside by a large south-facing window or glass door. In warmer regions, sow in containers in sunny spots and bring the plants indoors at night, so they don't suffer from large temperature swings. Stress-free plants will produce healthier crops.

Start off
Start peppers in late winter, sowing two to three seeds per small pot at a depth of ½cm (¼in). Leave the pots in a warm place, such as an airing cupboard, until the seeds germinate. When seedlings appear, move the pots to a sunny spot and thin to one seedling per pot.

When roots appear through the drainage hole, transfer the plants into larger pots filled with a mix of at least 70/30 topsoil and compost. Pinch out the growing tips of chillies once they are around 20–25cm (8–10in) tall for bushier growth. The final container for your peppers should be at least 5 litres (1 gallon) in volume (roughly 23cm/9in in diameter and 18cm/7in deep). Ideally, you should be able to pick it up and move it inside easily. Add two handfuls of compost to the base of the pot and fill with topsoil when you transplant the pepper seedlings. If you don't have compost, you can fill the pot with soil, and mix in some coffee grounds and wood ash.

PEPPER SEEDS ARE OFTEN
A WASTE PRODUCT LEFT OVER
FROM COOKING, SO THEY'RE
EASILY OBTAINED FROM
FRIENDS OR NEIGHBOURS.

Grow on
Water peppers thoroughly twice a week when the weather is dry. Once flowers appear, give the plants a liquid feed every two weeks to promote healthy growth. I use homemade comfrey feed (see pp74–75) because it is high in potash, which is necessary to produce fruit. Unless plants are overloaded, you won't need to support them with stakes or canes. For spicier chillies, water less often when the fruits appear.

Harvest
Peppers can be harvested when green or red. Pick them when green to encourage more fruits to form, or red for a sweeter taste. I pick my chillies when green, as they don't always ripen in the cool Welsh climate.

Chillies ripen at different rates, so both green and red (or yellow and purple) fruits can appear on the same plant at the same time. If you like your chillies hot, leave them as long as possible before harvesting, so the heat intensifies.

You can dry chillies (including seeds) for storage by hanging them up by the stems with the fruit attached in a dry, airy place. Alternatively, thread individual chillies on a piece of cotton or thin string and hang indoors until fully dried – they look amazing.

Saving seed
Extract seeds from ripe peppers and (with care) from chillies, then spread them out and place them on a sunny windowsill to dry out thoroughly. Store in an old envelope in a dark place so that you can sow them the following year.

I pick my sweet peppers when they're green (*top*), but in warm locations you may want to wait until they (*bottom left*) or your chillies (*bottom right*) ripen to their final colour.

TIP

When collecting seeds from chillies, wear gloves or wash your hands thoroughly afterwards to avoid a burning sensation when you touch your lips or eyes.

ALLIUMS

In this large group of flowering plants, the three key edible ones to focus on are garlic, leeks, and onions. You do need a little patience to grow leeks, but onions and garlic couldn't be more straightforward.

Alliums don't require huge quantities of nutrients and prefer light soil. Grow them in beds mulched with a 5cm (2in) layer of compost each year, or in large containers filled with an 80/20 mix of soil and compost.

a. GARLIC

Of the three recommended crops, versatile and strong-tasting garlic is the most suitable for growing in larger containers such as buckets, tubs, and tyre planters. If you are new to growing, planting this straightforward and undemanding crop is a great place to start your journey.

Start off
To ensure success, plant individual cloves 15cm (6in) apart and 5cm (2in) deep in their final position from mid- to late autumn. By late winter you should see little shoots, but there is no need to protect them from frost or snow. Garlic is remarkably hardy.

Grow on
Water garlic during periods of warm, dry weather to prevent the plant bolting and producing a flower at the top of the stalk.

Harvest
By early summer or once 50 per cent of the leaves have turned brown, the plants can be gently uprooted. Tie in bunches of 8–10 bulbs and hang up in a light, airy spot for a few weeks to dry out fully. Store the bulbs in a cool, dry place.

b. LEEKS

Leeks take a while to germinate and are slow growing, but the upside is that you can leave them in the ground until you need them.

Start off
In mid-spring, sow at least one row of leek seeds 1cm (½in) deep at the end of a raised bed or in a large container. You can expect at least 20 plants from a 30cm (12in) row, sown thickly. If you want additional rows, space them 15cm (6in) apart.

Grow on
Once leeks are approaching pencil thickness, they need to be transplanted into their permanent positions. Gently lift clumps of leek seedlings, plunge the roots in a bucket of water, and gently pull the plants apart. Use a stick to create holes 15–20cm (6–8in) deep and 20cm (8in) apart in your raised bed or a container of compost. Next, drop one seedling into the hole – you may need to push it down slightly so it reaches the base. Don't worry if only a bit of leaf is above ground. Pour water into the hole and let some of the soil fall back in, but don't allow the hole to fill completely – the stems will swell more if there's less resistance against them.

Harvest
In late autumn, you can finally begin harvesting and lifting leeks, possibly until early spring. Use a fork to loosen the soil and gently pull the leeks out of the ground at the same time, so that you don't damage the stems.

SAVE GARLIC CLOVES

The "seeds" of garlic are the individual cloves from the harvested bulbs. Set aside as many cloves as you need and keep them in a small cardboard box in a cool, dark place. The storage period for garlic is brief because you plant the cloves in the year you saved them. If you harvested your garlic in midsummer, for example, you can plant saved cloves in mid- to late autumn.

ONIONS FROM SETS

Sets are simply immature onions and will be ready for harvest sooner than an onion grown from seed.

Start off

In early spring, plant onion sets individually in newspaper pots filled with compost. Push the sets halfway down into the compost, place the pots on a windowsill, and keep them watered. Alternatively, in mid-spring, plant sets directly into the ground. Push half of each set beneath the soil and space at 10cm (4in) intervals. You may want to cover the sets with sticks until they have firmly rooted to prevent birds pulling them up.

Grow on

If you started your onions in pots, transplant them outside once the seedlings are around 7–10cm (2¾–4in) tall, allowing 10cm (4in) between sets.

Harvest

In late summer, when the leaves begin to yellow and the bulbs are firm, pull up the onions. You can use

them straightaway, or dry them before storing them. To dry onions, bring your harvested onions inside and lay them on newspaper in a cool, airy place, such as a shed or garage. Leave them in place until the foliage turns brown – this could take a month or two. You can leave them outside during dry weather to speed up the process, but they will rot if left in the rain. Tie the dried onions in bunches and use when needed.

ONIONS FROM SEED

I always grow onions from sets but seeds are easy to source.

Start off

Sow onion seeds in seed cell modules in late winter. Fill the modules with compost and sow three to four seeds 1cm (½in) deep per module. Place on a warm, sunny windowsill and keep watered. Onion seeds only remain viable for a couple of years.

Grow on

Once the seedlings are around 5–7cm (2–2¾in) tall, transplant into the ground, spacing the clumps 10cm (4in) apart. You don't need to thin because the onion bulbs will push away from each other as they grow.

Harvest

As for onions from sets.

I plant onion sets (immature onions, *above*) rather than seed. Leave onions to dry before storing (*opposite*).

KEEP YOUR SOIL HEALTHY FOREVER

Respect your soil, keep it healthy, and you will be rewarded with strong plants that produce abundant crops. A healthy soil is chemical-free, has plenty of organic matter, and is home to earthworms and millions of beneficial microbes.

Building up healthy soil takes a little time and effort, but your plants will be better equipped to take up nutrients, cope with periods of low rainfall, and fend off diseases.

Mulching your growing area with compost at least once a year will improve the structure of the soil and replenish the nutrients taken out by hungry food crops. Spread at least 3cm (1¼in) and ideally 5cm (2in) of compost (or well-rotted manure) over the entire surface sometime in the autumn. There's no need to dig it in. Then add a layer or two of cardboard (or at least 10 layers of newspaper) to keep weeds down.

Remove what's left of the cardboard in spring and sow and/or transplant directly into the beds. Where produce is growing over the winter, apply the compost mulch as soon as you have harvested the crop then sow and/or transplant straight into it. If you are just starting out on the growing journey and don't have enough compost, apply homemade liquid feed (see p73) and use the chop-and-drop method below.

CHOP AND DROP

Use this quick and easy way to improve your soil whenever possible. Simply harvest your produce, cut off the excess leafy plant material, chop or tear into small pieces, and scatter this back over the growing area to return nutrients to the soil. When I harvest cabbages, for example, I pull up the plant, strip off the loose outer leaves, and return the torn-up pieces to the surface. Plant material placed on soil in a thin layer will break down more quickly than in a compost bin, and there are other free sources of nutrient-rich materials you can add. Chopped-up comfrey, nettle, dock leaves, and chemical-free lawn clippings can all be lightly spread over the surface. When they have decomposed, just add some more.

SOIL IN CONTAINERS

When growing potatoes or other tubers, return the soil to the container after harvesting, but with a few large handfuls of homemade compost mixed in to top up nutrient levels. Containers planted with other food crops, such as legumes and salads, can just be top-dressed with compost when you have harvested the produce. Container crops will also benefit from home-made liquid feed because the low volume of soil they are growing in won't support as many beneficial micro-organisms as the larger areas of soil in raised beds.

LOOKING AHEAD

Once you have enough home-made compost to apply an annual mulch to the soil in your beds and are practising chop and drop, there is no need to carry on applying liquid feed. When you've been growing your own produce for a couple of years, it's a good idea to rotate crops grown in raised beds or the ground to preserve nutrients and guard against soil-borne disease. This simply means changing the position of different types of vegetables so they don't grow in the same place every year, and I explain this practice in detail on pp168–169.

To improve the soil health, spread a layer of compost on your beds in autumn (*top left*), mulch overwintering crops, here spinach, with dry autumn leaves (*below*), and use the chop-and-drop method to add nutrients when harvesting (*top right*).

FIGHT PESTS AND DISEASES

THERE IS NO SUCH THING AS A PEST- OR DISEASE-FREE GARDEN, BUT I EMPLOY A RANGE OF ORGANIC STRATEGIES TO KEEP PLANTS HEALTHY AND MINIMIZE PROBLEMS.

SLUGS AND SNAILS

These slimy creatures show your vegetables no mercy and – worst-case scenario – will happily munch a whole row of seedlings overnight. My approach to slug and snail issues is to focus my efforts on prevention and follow up with direct action.

DEFENSIVE ACTION

Controlling slug and snail populations isn't difficult, and although below I specifically mention slugs, the simple techniques I use apply equally to snails. Defensive action works best in the long term, which means creating barriers between slugs and your crops, and making sure plants are better able to fend for themselves when they are transplanted outside. And remember that slugs will always home in on tender seedlings, which they find irresistible, so give these the most protection.

It's always better to be safe than sorry, as the saying goes, when preparing for the start of the growing season. The following simple measures will make both your growing area and your plants less attractive to slugs, and ensure fewer problems in the long term.

START SEEDLINGS IN MODULES

Slugs head for small seedlings because these have the juiciest leaves and stems. By starting seedlings in modules under cover, they quickly progress through the young, tender stage. Once leaves and stems develop, the foliage toughens up and becomes much less palatable to slugs when planted outside. If you are concerned about slugs or currently have a large slug population, sow your brassica and salad seeds in modules on your windowsill and grow them on until the plants are quite large before transplanting. A lettuce with 10 leaves will withstand attack by slugs for considerably longer than a very young plant with just two or three leaves.

CLEAR THE GROUND OF WEEDS

Slugs prefer to hide or move around under cover, often lurking beneath the foliage of weeds. Try to keep on top of weeds in and around your growing area so the slugs aren't encouraged to emerge from their cover at night and make straight for your vegetables. Use the tips on p124 to control weeds.

KEEP GRASS SHORT

Slugs love hiding in long grass during the day because it is relatively dark and damp. Take precautions by keeping any grass paths around or between beds as short as possible, and avoid problems by trimming the edges regularly.

REMOVE HIDING PLACES

As well as long grass, slugs like to hide away under discarded pieces of wood or old pots left on the ground. I see plenty of these objects on unkempt allotment plots, or you might have a few lying around in your own garden. When you clear the ground of these enticing slug habitats, the cardboard slug trap I describe overleaf will be much more effective.

CREATE BARRIERS

Slugs hate sliding over spiky material so construct a barrier around your crops made from 30cm (12in) sections of bramble stems or other thorny twigs. Brambles are fairly easy to source for free, even in urban areas – try waste ground or common land. I've had great results using this method, but I'd advise you to wear thick gloves when making mini-fences as it can be a prickly process!

Slugs (*top left, top right*) and snails (*below*) come in all shapes and sizes. Remove hiding places by clearing weeds and cutting your grass.

ON THE ATTACK

After following the prevention strategy to reduce slug and snail numbers, you will inevitably see some of these hungry creatures still looking for food around your growing area. I'm not sure there is such a thing as a slug-free garden. The following methods are, in my experience, the most effective ways of dealing with these pests head on, without having to spend any money.

TORCH AND BUCKET

On a warm, humid night, go out to your growing area with a torch and bucket and check your plants. Look underneath leaves, between plants, and on the stems, then pick off any slugs you see and drop them in the bucket. Either squish the slugs or, when you've collected as many as you can, fill the bucket with water, cover, and leave for a week. You can then empty the contents onto your compost heap. A torchlight foray is one of the most effective methods of reducing the local slug population, and you'll notice a big drop in numbers if you go out collecting three nights in a row.

CARDBOARD TRAP

The best thing about this simple trap is its success rate. In the afternoon, place sheets of cardboard on the ground around your growing area. Return early the next morning, turn over the sheets, then pick off and destroy any slugs you find hiding there. This works particularly well after a rain shower: you will be laying cardboard over damp ground, which is where slugs like to be. Don't put down cardboard sheets if you are likely to be away or can't get to your garden for a day or two because you will, in effect, be creating a slug shelter right next to your plants.

WHEN PREVENTATIVE MEASURES HAVE BEEN PUT IN PLACE, I TAKE DIRECT ACTION AS AND WHEN NECESSARY

1

2

SLUG PUB

This simple trap makes good use of leftover (or stale) beer or lager, which slugs seem to have acquired a taste for. Source it from friends and family, or you might find an out-of-date can at the back of your cupboard. A slug pub is very quick to assemble – you can make one and put it out the same evening. Only use this when you are suffering severe damage and don't have time to use the torch and bucket or cardboard trap methods.

1. Sink a clean glass jar or a large, rinsed-out yoghurt pot in the ground with its rim about 3cm (1¼in) above the soil surface so you don't trap any beetles. Place it around 30cm (12in) away from any young plants that are showing signs of slug damage.

2. Pour enough beer or lager in the jar to fill it around three-quarters full.

3. In case of rain, add a small roof made from a piece of wood or cardboard held up by four sticks pushed into the earth around the jar. Then leave your trap in place overnight.

4. Take off the roof the following morning, check if any slugs have drowned in the liquid, fish them out, and dispose of them.

3

4

TIP

Don't put out a slug pub if there is no sign of damage to foliage, or you will lure slugs straight to your vegetables.

COMMON PESTS

The crops you choose to grow will be irresistible to a range
of hungry visitors from birds to caterpillars, but these free
or low-cost measures are highly effective in deterring them.

a. Birds
Although happy to share my soft fruit harvest with
birds (which eat insect pests), I do stop them getting
more than their fair share. Always harvest your fruit,
especially red berries, which blackbirds love, as soon
as it is ripe. Sparrows will also peck at chard leaves.
Homemade bird-scarers using discarded CDs or
DVDs will last for years. Thread seven to eight discs
on a piece of string and tie high up where they can
move freely in a sunny, exposed location. As the wind
blows, the reflected light from the discs bounces
around and unsettles the birds. You could also tie
strips of tin foil to a tall stick to create the same effect
and move it around the garden every few days.

b. Aphids (blackfly and greenfly)
By feeding on the sap of young plants and their fresh
growing tips, these pests can reduce vigour and
harvests. Simply rub them off by hand and cut out
badly infected leaves or stem tips.

c. Rabbits
These destructive pests can decimate vegetables in
rural plots and a dog is possibly the best deterrent.
If that isn't an option, make sure any perimeter fences
or walls are well maintained with no small rabbit-sized
gaps. Always reduce cover around your growing area
by removing piles of material and keeping grass low.
Live traps also work well and are often readily
available to borrow – ask around online or at your
local gardening club.

d. Cabbage root fly
Protect young brassicas from these maggots by
making 10cm- (4in-) wide cardboard "collars". Make
a hole in the middle for the stem, and a slit from the
hole to the outside edge, then fit around the base.

e. Cabbage whites
The small cabbage white butterfly's eggs turn into
pale green caterpillars; the large variety's hatch
into black, yellow, and green ones. Both can wreak
havoc on your brassicas. As soon as you spot the
white butterflies, check the tops and undersides
of leaves for eggs and caterpillars. Squish both.

f. Carrot root fly
Carrot tops turning yellow prematurely usually mean
this fly's larvae are feeding on the roots. Sourcing fine
mesh (or repurposing an old bed sheet) and erecting
a cover at least 60cm (2ft) above ground level will
prevent any female flies laying eggs, or you might
find seeds of resistant carrot varieties at seed swaps.
Planting carrots among other crops such as lettuce,
spinach, and beetroot, and growing all of them close
together, also reduces potential damage.

g. Flea beetle
Small round holes in the leaves of plants and seedlings,
particularly rocket and turnip, is a sign of flea beetle.
Rocket leaves won't look appealing but are still edible,
and although my turnips have never been completely
free of flea beetle, the harvests are fantastic. Regular
watering can help keep beetle numbers down.

h. Leaf miners
I tolerate the holes in chard, beetroot, and lettuce
leaves caused by these burrowing larvae, but you
can always remove any heavily affected leaves.

i. Pea moth
You won't know this pest has struck until you find
its caterpillars inside pea pods. Sow peas in early
summer before the moth's egg-laying stage and
get a later harvest, or grow mangetout instead.

COMMON DISEASES AND DISORDERS

Tiny organisms in the air and in rainwater can infect the soil
and weaken your crops. The first step is to identify the problem,
then follow the advice I give below to keep it under control.

a. Blight (early and late)

The early blight fungus strikes in wet, warm summers, and the symptoms are brown spots on the leaves of potatoes and tomatoes – it usually infects the lower leaves first then works its way up. Cut off and burn any affected foliage; I tend to remove all tomato leaves below the first truss of fruit because these leaves are more vulnerable to attack. Wet weather after midsummer can result in late blight affecting potatoes (and sometimes tomatoes). Its spread is more rapid than that of early blight, and infected stems and leaves quickly turn brown and rot. Act fast by removing all the foliage, then harvest the tubers. Grow potatoes in a different place the following year to avoid reoccurrences.

b. Blossom end rot

This disorder usually afflicts tomatoes, but peppers and courgettes might also be affected. The main symptom is an unappealing black circle at the base of the fruit. Erratic watering is often the cause, so keep to a strict watering regime. I'd suggest watering tomato plants deeply every other day during sunny, hot weather, every three days during overcast weather, and two to three days after rainfall (unless it is exceptionally hot).

c. Clubroot

Brassica roots affected by this fungal disease become distorted and weakened, which results in weak growth and poor crops. If it occurs, rotate your crops (see

a

b

c

pp168–169) and don't accept brassica plants grown on plots that have also been infected by the disease. If you know ground is infected, you can reduce the effects of clubroot by starting seeds in pots filled with homemade compost and only transplanting once seedlings are as large as possible.

d. Damping off
This disease affects young seedlings, which seem to wilt although they have had plenty of water and subsequently die. You can avoid damping off by never overwatering or overcrowding young seedlings. Always allow plenty of space between them for good air circulation. See p120 for more information.

e. Mildew (powdery and downy)
White, powdery patches (*see image, below*) on leaves indicate the presence of this disease, which usually occurs in the later stages of the growing season and thrives in dry conditions. It tends to affect squashes, so water these plants regularly and make sure you remove and burn infected leaves to help contain the spread.

Less common than powdery mildew, downy mildew strikes in wet weather and is most likely to affect the foliage of brassicas, rhubarb, and lettuce. The leaf blotches range from brown to pale green, and infected leaves should be removed promptly. Avoid overwatering, try not to splash water on the leaves of susceptible crops, and check foliage regularly.

f. Rust
Alliums can suffer from this disease, which is one of the easiest to recognize. Affected leek stems, for example, will be covered in small orange spots that look just like rust on a piece of metal, hence the name. The disease will reduce the vigour of the plant rather than kill it, so remove infected leaves promptly to stop the spread and keep your plants productive.

g. Scab
Primarily a disease of potato tubers, this fungus can also infect parsnips and beetroot, especially in hot summers with low rainfall. You won't know you have it until you harvest some of the crop and see rough brown patches on the skin. Reduce the risk by watering regularly when tubers and roots are developing. In very dry periods, water potatoes in the early morning and then mulch with cardboard to stop the moisture from evaporating.

d

e

f

g

CROP ROTATION

Growing the same crops in the same positions year on year can encourage pests to attack your plants and diseases to build up in the soil. Avoid serious problems and encourage healthy soil and plant growth by rotating groups of vegetables with similar requirements.

If you are new to growing food and have created beds from scratch, get into the habit of rotating your crops to maintain the health of your soil and deter pests. The key to crop rotation is to divide up your growing area and plant together vegetables that enjoy similar conditions and suffer from similar pests, such as brassicas. The following year, grow brassicas in an area where another group – for example, legumes (beans and peas) – was planted. In this scenario, moving brassicas to a new position not only confuses the pests that attack them but also allows these plants to take advantage of the nitrogen-rich soil left by the legume roots.

THREE-YEAR PLAN

I suggest starting off with a very simple crop rotation plan that divides vegetables into the following three growing sections that rotate over three years:

Section one roots and potatoes
Section two brassicas and salad leaves
Section three legumes, alliums, and squash

In this rotation scheme, each of the three groups of vegetables moves to a different section every year for three years. In year four, the crops go back to their original positions. The diagram (*right*) shows the crops growing in different beds, but you can simply divide your growing space into three sections and follow exactly the same plan.

YEAR ONE

YEAR TWO

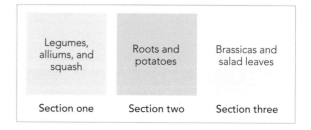

YEAR THREE

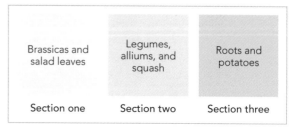

REDUCING ROTATION

When you have been growing your own food for two to three years, you can reduce the rotation of certain crops that suffer less than others from pests and diseases. These are: lettuces (and salad leaves), runner beans, and squash. Growing these crops in the same position can save time, and knowing they will occupy a designated space allows you to plan other crops around them. Not having to rotate runner beans, for example, means that if you built a sturdy structure for them, you can keep this in place over winter and reuse it the following year.

By year three of growing your own food, you should have a consistent supply of homemade compost (see p62) as well as very healthy soil. At this point, you have the option of rotating other vegetables less, but only if you haven't experienced any serious crop issues, such as carrot root fly, or potato blight in a wet summer. If either of these has taken hold then you must grow the affected crops in a different position the following year.

ROTATION IS A TRADITIONAL PRACTICE THAT GETS RESULTS WITHOUT THE NEED FOR CHEMICAL CONTROLS

ATTRACTING BENEFICIAL INSECTS

I enjoy allowing some annual vegetables to produce flowers so their pollen attracts beneficial insects into the garden. Florence fennel (*see below*) – a vegetable I haven't included in this book but that you can certainly explore growing later down the line – has the most gorgeous flowers that will entice pollinators, while leek flower heads are very popular with bumblebees and look fantastic. When beneficial insects visit your growing area, they will seek out other crops, such as courgettes, that produce separate male and female flowers. When the insects visit both types of flowers, pollination takes place and the plants start to produce a crop.

LOOKING AHEAD

ONCE YOU HAVE DEVELOPED A SETUP
THAT IS SUCCESSFULLY PRODUCING
FOOD FOR FREE, THINK ABOUT
EXPANDING YOUR OPERATION

MAKE AN INVESTMENT

You don't need money to grow a reasonable quantity of food, but you might want to expand or invest in some tools. Selling home-grown plants and produce is a great way to raise the necessary funds.

In your first year of growing, progress may seem slow, but as soon as you have achieved that first harvest, you will have completed the hardest part of all. Successful growing is simply a question of learning the right techniques, finding out where to source materials, having good ideas, and acting on them.

Also, the more connections you make with like-minded people, the faster things fall into place and the more enjoyable the process becomes. However, it's vital that you don't put pressure on yourself by setting unrealistic goals. I like to split tasks and projects up into bite-sized, achievable chunks so that I get those "quick wins" that keep me highly motivated.

When you have spent a reasonable amount of time perfecting your growing skills, you may be tempted to increase your yields and make life easier by buying a piece of kit, such as netting or a new spade. Some of the following items might be on your wish list:

- seeds of certain vegetable varieties
- more tools
- a wheelbarrow
- watering cans
- a mini polytunnel or cold frame
- protective netting – for extra protection from birds
- a water butt or industrial-sized water tank

ESTABLISH PRIORITIES

A good strategy is to prioritize three items that will make it easier to grow food in the future. When you have established your top three items, calculate the price of each, add up the total outlay, and make a note of it. Once you start receiving income from home-grown produce, record it straightaway, so you can keep track of how close you are to your target. I also like to put aside a proportion of my income from selling. This goes into a small emergency fund for last-minute seeds or necessary tool repairs. In the meantime, well-maintained, second-hand kit, such as a sturdy wheelbarrow, might come to your attention. Don't miss out: it will definitely be worth buying to cut costs. I often find that the best places to look are second-hand shops where you can get fantastic deals on high-quality tools.

A GOOD INVESTMENT

Water is vital for growing food so I would recommend that out of the whole list you prioritize an intermediate bulk container (IBC) tank. You might be able to source a used one from a local industrial site, and delivery to your growing area doesn't usually cost much extra. Some tanks may have held toxic liquid and contain harmful residues, so always check before you buy. Once you've taken delivery, rinse out the tank thoroughly, then attach it to a downpipe.

A single 1,000-litre (220-gallon) IBC tank full of rainwater is a priceless resource for any gardener. There is a great sense of security in knowing that you can keep your crops watered, especially at the height of the growing season in warm, dry summers. The tank will last for many years.

Sell your produce in order to fund tool and equipment purchases. Wheelbarrows and water storage tanks, such as butts, are among my top priorities.

If you have a glut of a particular crop, such as squash (*top left*), put some of your surplus up for sale or use it for bartering. Seedlings of plants that are easy to grow, such as kale (*above*), runner beans (*left*), or strawberries (*top right*), are also good options.

WHAT TO SELL

When deciding what to sell to raise funds, I find it helpful to break produce down into two categories: plant-based and harvest-based. In the first couple of years, I'd strongly recommend you focus on the plant-based category.

Selling plants you have grown from seed or propagated from cuttings is much more lucrative than selling your harvest, which has taken time and effort to grow. Seed saved from your own plants costs nothing, and neither do cuttings grown in homemade compost.

PLANT-BASED PRODUCE

The first priority is to identify which plants are most likely to sell of those you can propagate or grow from seed. Another important consideration is which of these require minimal effort. Below are five plants or plant groups I would recommend you prioritize.

Rosemary

People love the scent of this herb, and you can take dozens of cuttings every year from one established plant (see pp96–97). Start selling when they are a year old or offer two-year-old plants for a higher price. Lemon verbena is another popular herb. As soon as anyone smells a leaf, they can't resist buying a plant.

Tomato plants

Tomato seeds are so numerous and so easy to save (see p149) that you are likely to end up with far more young plants than you need. A great strategy is to get ahead by sowing seeds in late winter. By the time they've developed into strong, young tomato plants, demand for these high-value items will be at its peak.

Brassica seedlings

These crops couldn't be easier to sow and grow, and can be sold in ready-to-transplant cardboard tubes. Kale is a great brassica to start with because the seeds are easy to source and you often get hundreds in a packet. It also grows well in containers, so pot some up for selling to people who have only a limited space for germinating seeds.

Strawberries

Having explained how simple these plants are to propagate (see pp108–109), I suggest you aim to produce between 40 and 50 plants a year from layers. This number will cover potential sales with enough left over to use as a bartering resource.

Pea and bean plants

Like brassicas, these are easy to grow in small pots or cardboard tubes, and runner beans, especially, are always popular. Simply save enough seed from home-grown crops for your own requirements, with some left over to grow on and sell. From 20 pea plants I managed to save over 700 seeds – not a bad result!

HARVEST-BASED PRODUCE

When you have a glut of a particular vegetable, I'd always recommend offering the excess for sale. If you plan to sell your fresh produce anyway, I'd suggest growing these highly productive crops. All have a good shelf life:

- courgettes and other squashes
- root vegetables
- rhubarb
- potatoes and Jerusalem artichokes
- onions, leeks, and garlic

PROTECTED SEEDS

Some new varieties of seed are protected by plant breeders' rights (PBR) and propagation is tightly controlled. Avoid any problems by growing heritage varieties and not naming varieties on plant labels.

HOW TO SELL

Before you start selling, you'll need somewhere to display your plants and produce, as well as a container for collecting money. You may not get it right first time, but you will soon find a system that works for you.

Although selling your own seasonal produce on a small-scale shouldn't be classed as a business, I would suggest contacting your local or parish council to make sure that you are complying with any local regulations. Also, if your venture is a runaway success, you might have to declare the income on your annual tax return!

PRICING AND PAYMENT

Deciding what to charge can be tricky, but I've found two pricing strategies that are widely accepted by local authorities. Asking people to follow a "pay what you feel" approach strikes a friendly tone and indicates a degree of trust. Most people place a high value on fresh, home-grown produce and pay accordingly. If you are getting less for your produce than you think it's worth, display a "suggested donation" for each item.

The easiest way to collect money is to use an honesty-box system, leaving a container with a lid alongside your produce into which people can drop their payment. Unfortunately, people aren't always honest (I know this from experience), but you can reduce losses by checking your money daily, and bringing your produce and collecting box indoors overnight.

You may want to sell plants, too. Not only does this generate more income than selling produce, but people are less likely to take plants without paying for them. Fresh tomatoes might well be stolen from your display, but bean seedlings will be of interest only to those who will take the time to care for them.

DISPLAYING YOUR WARES

I recommend creating a small stand or produce box out of reclaimed wood, with an overhang to protect plants from the rain. You could also put produce in a weatherproof box with a lid. Plants don't have to be undercover, but will look much less appealing if they are dripping water after a heavy shower.

Here are my top three priorities when selling from your stand:
- Always display freshly harvested items and replace any produce that has a short shelf life, such as salads, after two days.
- Brush soil from your vegetables, but to maintain freshness, don't wash them.
- Sell only plants that are 100 per cent healthy with no visible defects.

KEEPING TRACK

This sounds obvious, but always note down everything you sell. This is valuable information that not only indicates what is popular in your particular area, but also allows you to adapt what you grow and propagate to suit local demand. Keeping track of income will also give you an idea of how much you've raised towards future projects or tools. I like to keep all the funds I raise in a glass jar for easy access.

SEASONAL VEG BOX

With enough growing space and about three years' experience under your belt, why not consider a small-scale veg-box scheme? Invite family, neighbours, and friends to invest in a weekly or fortnightly delivery of a box of seasonal vegetables at a set price. Start by supplying during summer and autumn, when you should have plenty of produce. Your income will be consistent – not always the case with honesty boxes!

I sell both produce and plants, using an honesty-box system to collect any payment.

TIP

Always take time over your display when offering home-grown produce for sale. People are more likely to stop and buy clean, freshly picked veg that is arranged attractively.

THE FIRST SWAP IS ALWAYS THE HARDEST (RATHER LIKE THE FIRST YEAR OF A NEW GARDEN), BUT THE SECOND ONE WILL BE MUCH EASIER

Make sure your display always contains a selection of your own produce at the seed or plant swap. For example, grow your own strawberry and tomato plants (*above*) and offer saved runner bean seeds (*left*).

SETTING UP A SEED OR PLANT SWAP

If you can't track down any local seed or plant swaps, the solution is to set up your own. It's bound to be a success because gardeners always have surplus seeds and plants, and can't resist getting their hands on even more.

A seed or plant swap is the perfect opportunity to pass on excess seed or plants from your own garden, and the event itself is a great opportunity to meet other growers and like-minded people. You might even come across those elusive plants you've been looking for. The ideal time to hold a seed swap is early spring, when everyone is thinking about what to sow but hasn't yet started in earnest; start your preparations straight after the new year. Plant swaps are most successful in mid- to late spring. Follow the steps below to set up your own local swap.

1. Contact local growers
Collaboration is the key to a successful seed or plant swap. Contact local allotment or gardening clubs to find out if anyone is interested in helping you to organize the swap. Aim for a group of three people – many hands make light work.

2. Find the location
Community centres, village and church halls, or allotment sites that have a large shelter make the best locations for swaps. Many will accept a donation in return for hosting the event, and you could always hold a raffle to raise money. It is also vital that there is good access to the site as well as parking nearby.

3. Seek donations of seed or plants
A friendly approach to local gardeners, garden centres, and seed suppliers will often get results. Most gardeners are happy to pass on seeds, and some suppliers will donate stock they can no longer sell. Local garden centres, too, may offer packets of seed from the previous season that are still viable, and they might even consider hosting the swap.

4. Invite speakers
Turn the swap into a small event by inviting people to give short 10-minute talks, or ask a passionate local expert to sit at a "help desk" and answer gardening questions. You can always show your appreciation with a small thank-you gift, such as a bottle of wine, after the event.

5. Spread the word
To maximize the attendance, choose a date and time on a weekend, such as a Saturday morning, and advertise the swap at least six weeks beforehand. Emailing local garden groups, printing off posters, and posting on local social media groups are all effective marketing strategies. You could even try asking a local radio station to promote the swap or interview you about it.

6. The big day!
Put similar types of seed in separate boxes for easy sorting and display them on tables. Remember to include a donations box so people who don't have seeds to swap can still participate. At least two of your team need to be around to help sort seeds donated on the day and to answer questions. Finally, show respect by leaving the location spotless.

7. Feedback
When everything has been cleared away, sit down with your team for a quick five- to 10-minute feedback session. Identify what went well and what didn't, so you can make next year's event an even bigger success.

WHEN TO EXPAND YOUR GROWING AREA

Growing your own food is a fantastic learning experience that will improve your skills and build your confidence. When your setup is working well and your crops are producing delicious yields, you might feel it's time to expand your operation.

Small, well-planned growing areas can be surprisingly productive and provide maximum harvests. Acquiring additional space, however, gives you the opportunity to be more adventurous with your choice of crops, as well as start new projects.

Armed with plenty of practical experience and the knowledge to tackle potential problems, the prospect of another blank canvas can be really exciting. But a word of warning: before taking on a new area, read through the statements below and note down your responses to each.

- You have spare time on your hands.
- You have maximized productivity using successional planting (see p42).
- You have established a good compost setup.
- You already have sufficient tools and water storage.
- You have been thinking about acquiring more space for a while.

ASSESSING A NEW SPACE

Once you can answer a confident "yes" to all the statements in the checklist, it's time to either expand in your own garden (building new raised beds, for example) or to find another location to grow in. Always bear in mind the SWAGA acronym (see p14) whenever you assess a new growing space. Its proximity to where you currently grow is particularly important because you may need to transport a couple of wheelbarrows of compost and move tools between locations. If you intend to grow in someone else's space, invite them to look at your current setup to reassure them there won't be a weed problem!

GROWING IN TWO LOCATIONS

Your own garden (or the space closest to your home) is the best place for quick-growing crops that need regular harvesting, such as salad leaves, kale, and root vegetables. An additional location will be ideal for those crops that require less maintenance, such as leeks, beans, and any perennials that you are growing. Compost and water are very heavy to transport, so prioritize creating one or more compost bins and increasing your water storage capacity. In your first year, begin filling your compost bins straightaway, and plant out perennials propagated from your own stock. I'd suggest waiting until the second year before you start growing annuals.

Growing food is all about taking lots of small steps, which very quickly add up to substantial yields. If you have raised some funds and access to the new location has been granted, why not get bulk compost delivered? This will give you a head start and the price is usually affordable.

MAKE MORE COMPOST

I'd advise holding off finding a new area for growing until you have a compost setup that is working at full capacity. Creating extra compost bins is a more efficient use of your time and current space because it will result in much greater productivity. Recycle pallets (see pp60–61) to make one or more more bins for free. Use local resources to fill them up and consider providing neighbours with containers (with lids) to fill with kitchen and garden waste.

In small and efficiently organized areas, there usually isn't scope for change. Another growing area will offer extra capacity and more flexibility.

A THREE-YEAR PLAN

This simple plan sets out the key stages on your journey of growing food for free. Take it one step at a time, enjoy learning new skills, and feel your confidence grow after every successful harvest.

How closely you decide to follow the three-year plan is up to you, but it is intended as a helpful checklist to maximize opportunities and minimize future problems. I have kept it short and simple, with page references that will take you directly to the relevant information or technique.

SIMPLE STEPS FOR SUCCESS

The first year of growing does require the greatest effort, but the rewards are huge. Remember to start by taking small steps and always split complex tasks into manageable sections. I cannot recommend this strategy enough: it's much more motivating to cross lots of small jobs off the list than one big one.

By the end of your third year, you will have a huge amount of growing experience under your belt and an efficient setup. I also believe that you will have the confidence to grow things without needing to look at a seed packet (or even this book!). However small your space, you will produce healthy and bountiful crops – and probably sooner than you might imagine. In the following pages, I offer a detailed journal of my own year of growing food for free.

YEAR ONE

1. Find a suitable growing space (pp14–15).

2. Get as much information as you can about your local climate and keep an eye on the weather forecasts (pp24–27).

3. Build a compost bin, source materials, and fill it as soon as possible (pp60–61 and pp64–67).

4. Source perennial herb cuttings and grow them on for future propagation and bartering (pp96–97).

5. Collect containers to grow crops in (pp30–33).

6. Build a raised bed and fill it up (pp36–39, *above*).

7. Make nettle feed to boost crops, and source comfrey plants (pp72–74).

8. Grow crops from dried beans and other kitchen staples and visit a local seed swap (pp80–83 and p84).

9. Establish a water-storage system (pp22–23).

10. Choose three vegetable crops that you can save seed from (pp81–82 and p129).

11. From summer onwards, source perennial plants from swaps or by bartering produce (p87).

YEAR TWO

1. Use your first batch of homemade compost as a mulch, adding a 3cm (1in) layer to beds and containers at the start of the year.

2. Rotate your annual crops (pp168–169).

3. If space allows, build another compost bin (or two), so that you have enough to use (pp60–61).

4. Build more raised beds and fill them up in stages, using the suggested methods (see pp38–39).

5. Take saved seeds to a seed swap (*above*); sow others to grow into seedlings for a plant swap (pp84–87).

6. Barter or sell some of the perennial herbs you've grown on from cuttings (pp96–97).

7. Consider increasing your water-storage capacity (p172).

8. Start to sell plants and produce to raise funds (pp172–177).

9. Save seeds from as many plants as possible (p128).

10. Grow on the perennial crops you sourced last year (pp88–115).

11. Mulch beds and containers with 3–5cm (1–2in) of homemade compost at the end of the year (p54).

YEAR THREE

1. If you have the space, create beds directly in the ground to maximize production (pp40–41).

2. Propagate from the perennial herbs you grew from cuttings to maintain your supply for bartering and selling (pp96–97).

3. Now you have established an efficient setup, consider looking for additional growing space towards the end of the year (pp18–19 and pp180–181).

4. Take care of mature perennial crops, such as strawberries and currants, so they maintain good yields.

5. Propagate strawberries from offsets and grow on for selling or bartering (pp108–109, *above*).

6. Start a journal to keep track of your successes and problems. Read this before every growing season and learn from these experiences (pp186–215).

7. Continue to mulch your containers and raised beds with around 5cm (2in) of compost every autumn (p54).

HUW'S JOURNAL

MY YEAR OF GROWING
FOOD FOR FREE

MARCH

Seeds swaps are a fantastic resource for thrifty gardeners – I only needed to offer a helping hand, and I received a great collection of seeds.

WEEK ONE

This week marked the start of my journey to grow food for free. Because I'm not a fan of starting a project without a plan of action, I spent a few days collecting thoughts and priorities. I concluded that compost, seeds, and containers were the three key areas to explore.

On Monday I began collecting compost materials even though I have yet to build a bin – these are just being piled up in a corner ready for when I source some wood. It is essential that when I do build the bin, I can fill it as soon as possible, so that I have compost to use this winter and next spring.

With seeds in mind, I went through the kitchen cupboard and found some dried peas. They might well be a couple of years old, so rather than waste time on a failed sowing, I'll use the germination test to see if they're viable.

For containers, I'm collecting anything I could use as a pot and storing it in a large cardboard box. So far there are a few tin cans and cardboard rolls, but some yogurt pots will imminently join the collection.

WEEK TWO

Wherever I go, I'm always on the lookout for discarded items, and managed to source some old tyres and empty compost bin bags from a couple of neighbours. But I really hit the jackpot at my local tennis club, where I came across a stack of three empty pallets. It turned out they were destined for the dump and the caretaker was happy for me to take them. I carefully split the pallets and had enough wood to not only build the compost bin, but also a pallet raised bed ready for filling.

I was looking forward to attending a local seed swap at the weekend, and was amazed at how open the organizers were for me to act as an extra pair of hands in exchange for some seeds. The result was a packet each of radish, lettuce, kale, leek, and swede seeds. The latter three are fantastic winter vegetables and can be sown and then planted out once I have set up a raised bed or two.

I've also made a fantastic arrangement with a couple of neighbours, which means I can borrow their garden tools whenever I need them. In exchange, they have access to my equipment whenever they wish.

> YOU CAN START COLLECTING COMPOST MATERIALS EVEN BEFORE YOU HAVE A COMPOST BIN, JUST PILE THEM UP IN A CORNER UNTIL YOUR BIN IS READY.

I prefer to grow crops in raised beds. This is one that I built using wood reclaimed from unwanted pallets I found at my local tennis club.

WEEK THREE

While visiting my neighbours for a coffee, I noticed they were throwing out some potatoes that had sprouted. I couldn't let these go to waste and asked if I could have them. Seeing how confused they looked, I explained I'd be planting these as seed potatoes to produce a crop. That really caught their interest and I now have a source of extra compost materials, as they offered to help me by giving me some of their vegetable scraps and grass clippings.

The germination test on the pea seeds turned out pretty well. A third of the seeds tested were viable and have germinated, which is good news. My plan is to sow and then grow on these peas and save some of them as seed this autumn. That will give me plenty of peas and pea shoots to harvest next year. For now, I am storing the seeds in a glass jar until I have somewhere to plant them.

Now spring has arrived, my priority is to get my perennials into the ground as soon as possible. On Friday I joined a local gardening social media group and posted a request for any spare edible perennial plants or cuttings and the response has been phenomenal.

WEEK FOUR

From the social media request post, I managed to obtain the following:
- 10 strawberry layers
- three small herb plants (rosemary, apple mint, and sage)
- a dozen gooseberry cuttings
- six Jerusalem artichoke tubers
- a rhubarb crown

The gardening community is remarkably supportive, especially of people who are just beginning their journey. The only downside was arranging and going to pickup points, but I know that the first year always requires the greatest upfront time investment. The rhubarb crown came in a large pot. After planting it out in the garden with some kitchen scraps at the base, I filled the pot with some topsoil (dug from the garden) and placed five of the gooseberry cuttings in it to root.

The gardener who donated the strawberry layers and rhubarb crown asked if I was after any other items. I mentioned I'd gladly accept any spare seeds they wouldn't get round to sowing, however old the packets might be. Turns out I will be getting a seed delivery in early April!

The strawberry layers have been heeled into the lawn for now. Heeling in is an easy way to keep bareroot plants alive when you don't yet have an opportunity to plant them because it stops the roots from drying out. You simply make a cut in the lawn and peel back the grass, tuck in the layers, and cover the roots back over.

Potatoes are easy to source for free, just see if anyone has any old ones in their kitchen cupboard – they don't have to be as sprouted as these ones, mind!

APRIL

Coffee grounds can be used as a mulch to improve soil nutrients when you are short of compost. If you have a coffee machine you'll have a reasonable supply, but you can ask in local cafés and coffee shops if you need more.

WEEK ONE

I am getting closer to filling the first compost bin, and now that the grass is beginning to grow, I am finding it far easier to source clippings. I've also recruited a couple of neighbours to save their vegetable scraps for me. The thing holding me back is the empty raised bed, but I had an exciting delivery over the weekend. Back in early March, I contacted a couple of local arborists and suggested that if they ever had spare woodchip from a clearance job, I would take it off their hands – now I'm the proud owner of a large pile of woodchip, which is the perfect material for the method I'm planning to use to fill my first raised bed.

The seed delivery arrived, and I was blown away with the contents! Many are still well within the expiry date, which means that next spring I'll have every reason to hit the ground running. I'm struggling with the lack of compost I have to work with, but next week I'll make a liquid feed to keep plants healthy in the short-term. Whenever I go into the local town, I take a container to collect free used coffee grounds from the cafés. I plan to spread about a centimetre over the raised bed as a mulch to add extra nutrients.

DON'T THROW OUT-OF-DATE SEEDS AWAY, JUST SOW THEM MORE THICKLY TO MAKE UP FOR THEIR REDUCED GERMINATION RATE.

WEEK TWO

After filling the raised bed and mulching with the coffee grounds, I kept back some topsoil to sow the dried peas – between three and four to a pot. I found a couple of empty ice-cream containers in the cupboard, and I put the pots of peas in these. I have also thickly sown lettuce seeds in another container (with holes in of course!) half-filled with topsoil and mixed with some of the coffee grounds. These seeds are now on a sunny windowsill and I must remember to keep them watered. Placing a couple of sheets of newspaper over the pots is a great way of reducing evaporation, and I will compost the sheets once seedlings appear.

I lined two large tyres with old compost bags that I'd made holes in because I wanted to get the strawberry layers planted as soon as possible. Strawberries don't require a nutrient-rich soil, so they are simply growing in topsoil and already look settled in their new home. I hope to propagate a few runners too in late summer, providing the plants stay happy!

Cardboard egg boxes are perfect containers for seedlings – the modules are the ideal size and can be transplanted directly into the ground with the seedlings.

WEEK THREE

The compost bin is getting full! In preparation for an overflowing compost bin, I created a second compost bay using more scrap wood. I'm amazed at how easy it is to source materials. A friendly smile and a simple explanation of what you plan to do with them go a long way. The best discovery this week was that a neighbour's daughter keeps a couple of guinea pigs and has a bucket to collect all the used bedding for me. This is hugely exciting because guinea-pig manure can be used straightaway in the garden and I now have a reliable source of extra material.

Because I wasn't expecting the arrival of guinea-pig manure, I already had a batch of liquid feed made from 50/50 grass clippings and nettles brewing in a bucket in the shed. I forgot to put the green material into an old cloth first, but that's not a problem as I can strain the solids from the liquid in a couple of weeks' time.

WEEK FOUR

The peas are growing well, but I feel they need another week before being transplanted. However, the lettuce seedlings have got a little out of hand, so I have potted these up into egg boxes and newspaper pots to grow on for a couple more weeks before transplanting outside.

I decided to add newspaper around the growing area and mulch with some of the woodchip to make a lovely pathway rather than slipping on grass! I then created a second raised bed out of some large logs I had at the end of the garden and love how rustic it looks. Because this bed is lower than the pallet bed, I plan to dedicate it to lettuces and salad leaves because they don't require much soil depth.

The week finished with an excuse for celebration – the first compost bin is full! I gave the contents a thorough mix and have covered the top with some cardboard to remind myself not to add any more material. To speed up the decomposing process, I will turn the bin once a month because I really want compost for a busy autumn of planting, propagation, and preparation. The way things are going, the second bin might even give me extra useable compost for next spring too.

Turning your compost once a month will speed up the decomposition process – I'm hoping my first batch will be ready by autumn.

MAY

WEEK ONE

The peas moved into their new home this week – the first raised bed! I was desperate to get something into it, and remembered to plant the peas on the north side of the bed, so that they won't cast a shadow over everything else. I used some twigs I found to act as initial support for them to begin growing up. It will be interesting to see how tall they grow because I have no idea which variety they are!

I strained the liquid feed into bottles using an old sheet. It smelled truly awful and I was glad to put the caps on! This feed is ready to be used (after dilution of course) and the pea seedlings will probably be the first in line for a drink. There is a local plant swap this weekend, and I plan to take a few bottles of the liquid feed plus a batch of lettuce seedlings to see what I can get in return.

Pea seedlings need supports to climb up when growing outside. I start with twigs or branches, adding longer sticks as the plants get taller.

Perennial plants, such as this **loganberry**, are fantastic crops to pick up at plant swaps. Use them to propagate more plants to take to future plant swaps.

WEEK THREE

Knowing I'll need another raised bed by the time the leek seedlings are ready to be transplanted, I had to get my act together and source some more pallets. I figured that the best thing to do was to speak to a couple of local builders who I have known for some time, and both were happy to drop off damaged pallets when they happened to be passing by. There isn't much call for damaged pallets in the building trade, but they are a fantastic resource for gardeners.

Now the rain is falling less frequently, my lack of water storage is going to be a major issue and needs to be rectified immediately. I rushed off to my local café with a large bag and asked them to save all their empty milk cartons. Two days later they had collected 20 and set them aside for me. In the meantime, I shortened the gutter downpipe so I could fit an old dustbin underneath. All I need now is some rain!

WEEK TWO

The plant swap was a success, even though I was late and there wasn't much choice left. I came away with a couple of cherry tomato seedlings, a tray of kale seedlings, and a loganberry plant. The loganberry will be perfect to grow along one of the fences and I have a beautiful variety of kale to grow – cavolo nero! I decided to lightly mulch the raised bed with grass clippings to continue adding extra nutrients to the soil, and the peas are looking lush and healthy. My only concern is that the raised bed is still looking fairly empty.

To help fill the space, I planted eight of the kale seedlings into the bed and sowed some radish around them in the hope of getting a crop before the kale gets too big and blocks out all the light. Leeks are the other seeds I completely forgot about and are one of the best crops for winter, so I left some space at the south end of the bed to sow a row of leek seeds. I'll transplant them into their final positions once the leek stems are the thickness of a pencil.

The sprouted potatoes I received back in March are growing well in a tyre filled with soil and vegetable scraps. In the area where I live, there is no shortage of used tyres and empty compost bags, so I "earthed up" the potatoes by stacking another tyre on top of the initial one. I filled it with a mix of grass clippings, topsoil, and guinea-pig bedding.

This year I started lettuce indoors and transplanted it out when the seedlings had developed, but you can also sow lettuce seeds directly into their final position.

WEEK FOUR

I had to transplant the lettuce seedlings into the salad bed as they had become a little too big for their containers and the leaves were beginning to lose colour. I planted them around 30cm (12in) apart to allow space for them to develop fully, and I made sure I welcomed them into their new home with a drench of liquid feed and a mulch of used coffee grounds.

I was travelling along a country road earlier this week and something caught my scavenging eye. I pulled over and saw a stack of at least 10 large animal-feed buckets shoved under a hedge. This was a clear example of littering, but I could see they would make excellent planters for beans, potatoes, tomatoes, herbs, perennial tubers, and soft fruit. I'd never have imagined that finding empty buckets could be so exciting!

There was no rain until Friday, so I had to use greywater from the kitchen sink to keep the seedlings and plants watered. Now the rain has fallen, I have a whole dustbin-full of water, as well as all the milk cartons filled to the top and ready for action. When the first drops fell, I felt a huge sense of relief.

I've adapted one of my gutter downpipes (*left*), so that I can place a large dustbin underneath to collect rainwater. Hopefully, this will mean that I always have enough water for my crops – even during dry spells.

JUNE

WEEK ONE

The highlight of this week was picking fresh pea shoots! I just couldn't help myself and really enjoyed them in a salad. The benefit of pinching out the shoots is that it creates a bushier and stronger pea plant, so at least I had a valid excuse for this sweet snack.

On a sourer note, it suddenly dawned on me that I hadn't sown the runner bean seeds. Time was running out and I was also low on cardboard tubes. Fortunately, back in April, I had dropped an empty bag at a local hotel and asked if they would be happy to collect the cardboard insides of loo rolls for me; turns out they had over 60 to give me! Of these, 25 are now sown with runner beans. I used the in-fill method to obtain some more soil (see p38), mixing in used coffee grounds and watering with diluted liquid feed.

I am very impressed with how everything is looking in the beds. The kale is growing rapidly and looking bigger by the day, and the potato plants are massive. It's also great to be harvesting some fresh herbs, although I know that this summer I must hold back from too much picking, and focus on taking herb cuttings to increase my bartering stock next year.

WEEK TWO

It was back to reality on Tuesday when I discovered the local slug family had found my prized lettuces and decided to do a taste test. The plants were fairly well developed, so weren't completely wiped out, but the slug attack will set them back a week or so. That day I cut several bramble stems and built a thorny fence around the lettuces, then set up a couple of slug pubs in the same area for good measure. This strategy has proved to be very effective and the lettuces have suffered only minimal further damage.

A very welcome delivery arrived on Friday afternoon. One of the local builders had saved some damaged pallets for me but now they're here I think I might have too many! Then again, I will need a third compost bin later in summer so all the pallets are sure to be repurposed.

I've been carefully monitoring the frost situation. We haven't had one for weeks, and the poor tomato seedlings desperately need transplanting. I removed some material from the initial compost pile to half-fill two of the animal feed buckets (having made drainage holes in the base first) and put in a mix of soil, coffee grounds, and grass clippings. The compost material will act as a fantastic sponge to retain water, and my tomato seedlings are now looking happy in their pots just outside the porch.

Slugs are definitely one of the most frustrating pests for gardeners. These are some I found on a hunt around my garden.

Pea shoots make a great snack
and can be added to salads.
When I have enough peas to
spare, I plan to sow them just
to harvest the shoots.

WEEK THREE

Work was quite demanding this week and I only had time for some watering. We haven't had any proper rainfall for a couple of weeks, so I've been saving greywater and using a lot of the rainwater to water the plants early in the morning. With a few hours to spare over the weekend, on Saturday I built another pallet bed and filled it using the in-fill method (see p38). I was planning to put the leek seedlings in it and I'm pleased it's ready and waiting for them.

WEEK FOUR

The long evenings are really helping me get on with growing, and I love being outside soaking up the last of the sun! This has felt like one of the most exciting and productive weeks so far on this journey. I began by sowing spinach in the salad bed and harvesting lots of tasty radish. The lettuces are also providing lots of delicious leaves, and I have been enjoying home-grown salads of lettuce, radish, and pea shoots.

I'm itching to harvest the potatoes, having felt a few tubers when I had a little dig around the roots. But they aren't quite ready yet and definitely need more time to grow, so I must be very patient. Even though we have finally had some more rain, I gave the potato plants a good watering.

Transplanting the leeks was so satisfying! I love how easy the process is, and there is something strangely enjoyable about dropping a seedling into a hole, but not needing to fill it with soil. The only downside with leeks is the wait. I can't harvest them until November/December, which seems so far away right now. I also made sure I transplanted the runner beans. It was a little disappointing that only roughly half germinated, but they are now planted in two more of those free animal-feed buckets. I need to find some long sticks to train them up. I've noticed that there is a lot of invasive bamboo growing in my local area, which would be perfect for the bean structures!

It's important to give transplanted seedlings, such as these leeks, a good water when you move them to their final growing spot.

JULY

WEEK ONE

I couldn't hold back any longer and just had to harvest the potatoes! I was really pleased with the yield and they taste delicious. We'll enjoy eating them for a while yet, and I made sure that I saved 10 potatoes for seed next year. I'm quietly confident that I'll manage to source even more seed potatoes to plant in March – I hope I'm right!

My salad bed is going from strength to strength. The lettuces are still cropping very well, and I'm watering more frequently to try and stop the plants bolting. Back in spring, I received some out-of-date beetroot seeds in the seed delivery from the kind local gardener. I'll sow them in the salad bed this weekend. I can eat the young leaves and hopefully I'll get some nice beetroot to enjoy, too.

Potatoes are a fantastic crop for a self-sufficient gardener. They are easy to grow and produce lots of tubers. Remember to save a few to plant next year.

WEEK TWO

Having got their roots down, the runner beans are flying up the bamboo supports. I gave them some liquid feed when I watered them, and I'm delighted that the heavy rain earlier in the week topped up my stored water levels. I now have more than 40 containers full of water and I have placed buckets under the shed roof to capture even more rainwater!

Herbs are a huge passion of mine and a big feature of our home cooking. The three herbs I planted back in March (rosemary, apple mint, and sage) are growing well, but I'd like to add more to the collection.

This week I managed to source cuttings of lemon verbena, lavender, peppermint, and thyme. They are all planted in yogurt pots except for the peppermint stems, which are in a jar of water.

My aunt and uncle have agreed to let me take divisions of some of their mature herb plants in the autumn. Soon I'll have marjoram, lemon balm, and chives, too. I'm so fired up about growing herbs that I plan to create a new raised bed this winter, just for perennial herbs.

WEEK THREE

My pea pods are now plump and ready to harvest. They are the best fresh garden snack ever – even better than strawberries, which I also harvested this week. In fact, I love fresh peas so much, none of the first harvest made it indoors! To my mind, enjoyment is the ultimate goal and reward for growing your own food, and although I really want to share these peas with others, I'm far too greedy.

I planted some Jerusalem artichoke tubers next to the rhubarb in April, and the rate of growth has been amazing. I have been mulching the tubers and rhubarb with plenty of coffee grounds and grass clippings, and both crops are in great condition. The liquid feed supplies have been running low lately, so I started another batch last week and used the last of the original batch on all the plants in the raised beds.

Yogurt pots are a great size for potting up cuttings from perennial herbs.

Mint cuttings are easy to propagate
– simply leave them in water until roots appear, then plant them up in compost-filled pots.

WEEK FOUR

The peppermint cuttings in water are already rooting! I can't help but get excited by the magic that happens when you put a stem in water and it grows roots. I will keep the mint in water for a couple more weeks before transplanting, and I'm keeping my eye on the other herb cuttings in pots, which are also looking good.

The salad bed is still fairly productive but I have noticed some of the lettuce leaves are starting to taste a little bitter. To reduce the bitterness, I've been steaming them lightly before eating, which works well. It will soon be time to start harvesting the spinach, so we'll have more fresh greens.

Around 20 of beetroot seeds germinated (out of at least 100), which was disappointing, so I plan to let these mature normally (no leaf picking) and will enjoy eating the roots this autumn. The sweet taste of fresh home-grown roots will always be far superior to any shop-bought beetroot.

On Saturday I turned the compost pile for the first time this month. As I worked, I noticed the centre looked and felt dry, so I added around five large recycled milk cartoons full of water. That should bring the moisture levels back up and also help speed up the composting process. I'm a little surprised by how healthy my veg plants are considering that I'm relying on just mulches and liquid feed to keep them going. In the autumn, when the contents of this bin are ready, I'll spread a layer of compost about 5cm (2in) thick over the raised beds and add the same amount to my pots. A lack of compost can limit how much you can sow and grow on, so once this precious resource is sorted, the toughest period of growing food for free will be behind me.

AUGUST

WEEK ONE

I had to accept that the lettuce had finally gone to seed. It tasted so bitter that even steaming wouldn't have made any difference. I decided to leave around six lettuce plants in the raised bed to flower and set seed, and then put the remainder in the compost.

Although being part of a small tool swap with kind neighbours has been working well, I would like to start reducing my dependence, on them and this week I had a real breakthrough on the tool scene. A long-overdue clear-out of my bedroom resulted in a pile of books, electrical items, and clothes that I no longer needed. After posting them on my local online swap shop with a request for gardening tools in return, I have sourced two forks, two large buckets, a spade, and a trowel in no time at all!

WEEK TWO

The leeks are growing well, and I think they always look their best and most luscious right after heavy rainfall. I gave them a light mulch of used coffee grounds and pulled out some of the weeds that have started appearing. I am very strict when it comes to weeding, and once a week I will go through the whole growing area to remove any that pop up. Visitors always comment on how tidy and weed-free the garden is and how much work it must involve. In fact, it's one of the easiest of garden tasks as long as you do it little and often. It's the secret to not feeling overwhelmed by these pesky plants.

The strawberry plants are sending out runners in every direction and seem determined to take over any of the surrounding bare ground. Some have even started rooting in the woodchip path. Over the last few months, I have been collecting empty tin cans from friends and family, and they are perfect for potting up the layers (the tiny plants at the end of the runners). I forgot to add drainage holes to the base of the first couple of cans, but that mistake was quickly rectified with a nail and hammer once I'd emptied out the contents. As an experiment, I filled the cans with around two-thirds half-decomposed material from the first compost bin and a third topsoil. I also potted up the mint cuttings in cans using the same mix.

Old tin cans are ideal for potting up cuttings, such as mint (*left*), or starting seeds in – just remember to make drainage holes in the bottom!

WEEK THREE

Most of this week was spent visiting family on the other side of the country. Although I was away for five days, I was confident that if I gave all the plants a deep watering the day before I left, they would be fine.

Fortunately, there were no disasters. Well-established plants, such as the kale and Jerusalem artichokes, would be able to survive for a few weeks without water, but the plants in pots were in need of another soak when the five days were up. It was nice to return to a handful of runner beans ready for harvesting as well as tomatoes at long last!

It is sad to see that the pea crop is slowing down, so I must hold back from too much snacking and let the pods mature and dry so I can save seed. Having discovered there are approximately six peas per pod, and that I have about 50 pods left on the plants, I won't be short of pea seeds next year to grow or take to seed swaps.

Jerusalem artichokes are some of the tallest plants in my garden. The stems shoot up quickly, but it's the tubers developing beneath the soil that you harvest.

Runner beans are the crop that keeps on giving – as long as there are still flowers visible, you know that you will continue to have beans to harvest.

WEEK FOUR

The runner beans have been a welcome addition to many evening meals and are really delicious. The tomatoes have replaced the peas as the garden snack, which is nothing to complain about, except that I'm having to be more patient on my forays to check the crop because there are fewer tomatoes available to eat in comparison to the abundance of peas!

On the theme of harvests, I decided to pick some kale leaves to add extra greens to Saturday's supper and they were delicious steamed with a sprig of rosemary. I am now getting a generous supply of spinach from the salad bed and love the fact that home-grown, free food is making an appearance in my meals every day of the week.

The days are shortening and autumn is on its way, but that means copious quantities of fallen leaves and plant material – all just begging to be composted. To my mind, the best way to prepare for autumn is to have enough free composting space, so with the second compost bin close to overflowing, I decided to create a third one. My goal is to fill this bin before winter hits. The contents of the initial compost bin should be ready in November, and I will use and store as much of that as possible, so the bin will be free for winter materials, such as vegetable scraps. I feel that a three-bin setup will keep me supplied with more than enough compost throughout the year to provide bountiful harvests every season.

SEPTEMBER

WEEK ONE

I don't like to think of myself as a hoarder, but since taking up the challenge of growing food for free, I can now see potential in so many random items around the garden. One of my favourites is an old metal sink that will make a perfect planter for herbs. I have also been storing materials for creating raised beds and now have quite a collection of stones, which I plan to use as edging for a perennial herb bed. My (arguably dangerous) philosophy is that the more material I can collect, the more choice I will have in terms of what I can create this winter.

Realizing I hadn't turned the first compost heap for at least six weeks, I decided to make this task my Friday morning workout. As I forked up piles of compost, I could see it was almost ready to use as a growing medium, and giving off a lovely, earthy, forest-floor smell, which is exactly what good compost should smell like.

Meanwhile, light mulches of grass clippings and used coffee grounds have been perfect short-term substitutes for home-made compost. I'll continue to use them next year, but only as occasional boosts because the soil will be very healthy once the compost has been added later in autumn.

WEEK TWO

The mint cuttings are nicely settled in their new homes, but I confess I'd forgotten about the other herbs I'd propagated. I put them outside in a shady corner that I hardly ever visit and I only remembered about them this week!

The herb cuttings were looking a little dry and unhappy, but at least I hadn't left them in a sunny position, where they would have certainly wilted and died. They had been planted in small pots and I'd usually pot them on into larger ones, but with winter approaching they will soon go dormant. Instead, I planted the cuttings in the space underneath the kale. I can lift them out in late winter and either pot them up ready for spring or plant them in their final growing position. The other advantage of this strategy is that I don't have to source more growing medium to fill pots and I needn't worry about the herb roots drying out.

The leeks have put on considerable growth over the last few weeks and I'm really excited about them. Leeks, along with kale, will be the two most productive crops with the biggest yields of this growing year. In the raised bed there are around 50 leeks, which will be harvested and relished this coming winter in soups, stews, pies, and caramelized in a little olive oil – there are so many ways to eat and enjoy them.

UNHEALTHY LOOKING PLANTS MAY JUST NEED A LITTLE WATER. IF THE SOIL IS DRY, TRY GIVING CROPS A THOROUGH WATER EARLY IN THE MORNING.

WEEK FOUR

The yield from the runner beans grown in the animal-feed buckets has been phenomenal. I was just about to blanch and freeze a big batch, but then thought I'd benefit more, in the long-term, by saving the seeds instead. Now I will have even more seeds to swap next spring!

All the strawberry layers are showing strong, lush green growth. This indicates that they now have a strong root system, so earlier this week I severed them from the mother plant. The strawberries I harvested over the summer were delicious – the blackbirds would definitely agree with that – so I will save six of the layers to plant out this winter. I'll put them in tyre planters to increase my yields and use the remaining rooted layers for swapping.

I've been enjoying regular harvests of spinach from the salad bed, but now the light levels and temperature have dropped, the supply is beginning to dwindle. The kale will soon replace spinach as my primary supply of greens, and I also have the option of fresh beetroot leaves now the roots have almost reached the optimum size. What I love about growing food is that there is something different to harvest and eat every season. And as soon as you start getting bored with one vegetable, another one comes along to replace it!

Wait until the pods on your pea plants have dried out and gone crinkly before you remove them to save the seed.

WEEK THREE

I have been checking on the pea pods for the last week, and this weekend they felt very crinkly and dry to the touch, so I collected them up in a container.

Once indoors, I extracted the peas and spread them out on a baking tray. I'll leave the tray on a sunny windowsill for a week before I store the dried peas in a glass jar somewhere dark. My guesstimate is that I have 300 pea seeds, which is far more than I initially expected to save! I plan to keep a third for growing next year and swap the rest for those seeds I really want.

It was satisfying to cut off the pea plants at the roots and compost the dried-out stems. They had grown to about 1.2m (4ft) in height up a simple homemade support I had made from wire attached to old fence posts.

I often pull out my pea plants with the dried pods attached – I find it easier to strip of the pods afterwards while sitting down.

OCTOBER

WEEK ONE

This week saw the first beetroot harvest and it was certainly worth the wait! Of the seeds I sowed, 20 or so plants grew to maturity, and they store well in the ground, so I will lift the roots as and when I plan to eat them between now and the start of November. I always recommend beetroot. It is one of the easiest vegetables to grow and always tastes noticeably better than anything you can buy in a shop. I can feel that the garden is now going into rest mode with everything slowing down for winter, but at least I will have time to tackle some big projects, such as creating new growing areas.

WEEK TWO

I don't think I have seen the sun for about three weeks! It has been wet, miserable, and cold – none of which is helping the runner bean pods to dry. Another drawback was the small amount of rust on the leek foliage, which is hardly surprising given the recent weather conditions. Rust is a fungal disease that must be contained if yields aren't to suffer, so I check the leek bed every other day and cut off any affected foliage. I always put diseased material in the household waste rather than the compost bin, so the chances of it spreading to contaminate other plants in the future is significantly reduced.

Garlic is easily one of my favourite vegetables, and for the past few months I've been putting my energies into sourcing some cloves for free in time to plant this autumn. My persistence paid off a couple of weeks ago when I was speaking to a neighbour, also a keen gardener. He was just about to put in an order for some garlic and I asked if he would consider getting a couple of bulbs for me and I'd offer something in return. All he requested was help to plant up his garlic patch, which we did on Saturday. For around half an hour's work I now have three bulbs' worth of garlic cloves to plant. One slight issue is that I don't have anywhere to plant them, so I must build a raised bed next week.

Although I left most of my beetroot in the ground, I couldn't help myself lifting a few to eat now.

I'm so pleased I managed to get hold of kale seedlings earlier in the year. It's such a productive crop that I'll be able to harvest it throughout the winter.

WEEK THREE

Back in the summer I received a broken pallet collar and had forgotten all about it. The only remedial work required was a couple of old nails hammered into one of the corners and I had an instant raised-bed structure for the garlic. I used a variation on the in-fill method, removing about 15cm (6in) of topsoil from the ground inside the frame and packing the space with woodchip. I then used the first of the homemade compost mixed 50/50 with topsoil to fill the bed. After allowing this to settle for a couple of days, I then planted 24 garlic cloves into the bed.

It has been a productive week because I also had the opportunity to divide some mature herbs in my aunt and uncle's garden. Amazingly, I came away with 12 divisions of a mix of lemon balm, chocolate mint, marjoram, thyme, lovage, and chives. One of each type of herb has been heeled into the ground ready for the perennial bed, and the rest have been potted up in various containers to take to a plant swap next year. I also managed to divide some of their rhubarb plants to add to my collection. Although it would have been better to wait until next month, rhubarb is a tough perennial, so my divisions should survive.

WEEK FOUR

Kale is now making a regular appearance on the dinner table, and I decided to make a fantastic batch of kale chips, which went down a treat. I have also been carefully monitoring the leeks for rust. Luckily, it has virtually disappeared and, judging by the size of the stems, the first will be ready to harvest in a matter of days.

A plant group missing from the garden this year is the cucurbits – courgettes, cucumbers, and squashes. Towards the end of this week, we were carving out a pumpkin for Halloween, and I suddenly had a eureka moment: why not save the seeds? Commercial pumpkins are grown on a large scale, which means they are very unlikely to cross-pollinate with any other type of squash, and so the seeds should grow true to type. I got carried away, carefully removing pumpkin seeds from the compost bin, and now have around 60 seeds drying on the windowsill. I plan to sow a batch in tin cans, grow them on and take them to a plant swap, each with the label "Halloween Carved Pumpkin Seedling". I hope we'll all get a successful harvest.

NOVEMBER

When using autumn leaves as a mulch, I cover them with sticks to weigh them down. This prevents the mulch blowing away in the wind.

WEEK ONE

One of my neighbours is a very talented gardener who grows the most beautiful bedding plants and roses. Having seen a few pots of herbs on the patio, I decided to challenge her to try growing edibles as well. When she asked what would grow well in a large terracotta pot, I recommended strawberries. She was delighted when I called round after work on Thursday with a gift of three strawberry plants I'd propagated earlier in the year. It felt great to give these away and encourage someone else to grow food for free, even on a tiny scale.

My wish for next year is to have enough home-grown material to enable me to supply free plants, cuttings, and seedlings to the local community on a regular basis. I want to encourage others to grow their own, or at least to take their first steps in the right direction.

I harvested the first couple of leeks this weekend, using a fork to gently loosen the ground around them. I was afraid that if I yanked too hard, I might snap the stem. The leeks are a fantastic length and the taste is something else. I caramelized them in melted butter and would happily eat a whole bowl of this and nothing else for supper.

WEEK TWO

Everywhere I look the ground is covered in autumn leaves. Whenever I go for a walk, I take a couple of hessian sacks sourced from a local artisan café. These once held coffee beans and are perfect for collecting and storing leaves. Neighbours are also happy for me to help myself to the leaves they sweep into big piles in their gardens because there is less for them to dispose of at the local green-waste centre.

I've been planning to mulch the garlic bed with a 5cm (2in) layer of leaves and leave them over the winter. The main drawback with leaves is that they blow around, so I lay a network of sticks over the top to keep them in place. At the end of the month, I will also mulch the strawberries and rhubarb with a layer of leaves, adding a layer of compost over the top to hold them down and provide a slow release of nutrients during the next growing season.

There are so many leaves to collect in autumn, that I often reach a point where have to stop myself gathering them – if I didn't I'd have more than I could use.

WEEK THREE

Last weekend I finally had to accept that there was no more room for leaves, unless I piled them up somewhere to make leaf mould. Life has been a bit manic recently and there isn't time to build a cage, so I compromised by dumping the leaves in the corner where the wall meets the fence. I weighed them down with empty compost bags and stones so that they wouldn't get blown away, and feel it's a good temporary solution. The pile may not look smart, but it means I can collect more leaves and add them to it.

The festive period is drawing near and close friends and family have been asking what's on my wish list. I haven't had much time to think, but I know I'd like an annual subscription to a particular gardening magazine that seems to have free packets of vegetable seeds attached to the front of almost every edition. To my mind, that's an amazing way to source a steady supply of seeds at no extra cost. And if I don't want or need some of the packets, I can exchange them for something else at the next seed swap.

WEEK FOUR

The compost in the first bin is calling to me, as it is now ready for use. There are a few clumps of partly composted material, but these have been added to the second bin. I used a spade to fill up a couple of buckets with compost, then carried them over to the raised beds and mulched the leeks and kale with a 5cm (2in) layer. I felt relieved that I'd finally produced some compost to put on these beds and knew I wouldn't have to think about adding more until next autumn.

The animal-feed buckets and tyre beds I'd already planted also got a 5cm (2in) layer of compost. Given this addition of organic matter, the fertility of the soil will increase and the plants will be given a boost.

More than two-thirds of the compost remains in the first bin, but because that is destined for future beds and tyre planters, there is a risk that I'll use it all up before spring. That would be a bad move because the compost in the second bin is still some way from being ready (hopefully I'll be able to use it in April, providing I keep turning it regularly).

Thinking ahead, I decided to fill two large boxes with compost and put them away in the shed. Now I know that when the time to sow seeds comes around, I'll have plenty of compost to fill the pots.

I HAD HOPED TO EMPTY MY FIRST COMPOST BIN BEFORE THE START OF WINTER, SO I COULD START FILLING IT UP AGAIN. HOWEVER, NOT EVERYTHING GOES TO PLAN, WHICH IS SOMETHING ALWAYS WORTH BEARING IN MIND.

DECEMBER

Use any soft fruit cuttings you can get hold of to propagate new plants. They are often discarded by gardeners, so try asking someone with a fruit bush for their prunings.

WEEK ONE

One of the best things about attending the seed swap earlier this year was being able to meet and chat to so many gardeners and allotmenteers in my community. I've always thought gardeners were a generous bunch of people, and when they heard I'd challenged myself to grow food for free, they were keen to help in any way they could.

One of the gardeners grew soft fruit and said I was welcome to any cuttings when the time came to prune her currant bushes and berries, so we exchanged numbers. I phoned her in November and this weekend I popped over and was told to help myself to as many as I wanted because she was going to burn any that were left. I came away with 50 cuttings – a mixture of currants, jostaberries, and gooseberries – which is far more than I need or have space for! I'm already growing on the cuttings of traditional green gooseberries that I sourced back in March, but those I took at the weekend were from a lovely red variety and I couldn't resist them.

The soft fruit cuttings have been heeled in the ground for the time being because I have yet to prepare a propagation area to grow them on. Given space restrictions, I will keep a maximum of eight of the cuttings when they have formed roots. The rest will be given away or I might even do some guerrilla gardening and plant them in nearby hedgerows.

During the late autumn and winter months, there is much less that you can do in the garden. I use this time to come up with new ideas for the following growing season.

WEEK TWO

This week I stuck to the theme of soft fruit and focused on the five gooseberry cuttings I planted in a pot back in spring. I carefully removed the tangled mass of roots from the pot and placed them in a bucket of rainwater for a couple of hours. This made it much easier to separate the plants and minimize any root damage in the process.

I decided to keep two of the cuttings and give away the other three to friends. I transplanted one into a tyre filled with soil and compost and then planted the other directly into the ground near the rhubarb patch. The whole garden is really starting to fill up with food crops and it should look even more productive in a year's time.

WEEK THREE

I've decided not to start any new gardening jobs until the New Year, although I'll still check on my plants and harvest winter veg. I'm planning to spend some cosy evenings gathering ideas for the garden that might be useful next year. The internet is the best source of free content, and I can't help getting sucked into watching gardening videos from all over the world. Having some time out and not worrying about the other tasks I need to do in the run-up to the festive period is blissful.

I'm using a notebook to jot down anything that grabs my attention – for example, growing salads in old guttering attached to a fence. Next year I can look through all the notes and sketches I made and pick out the ideas that excite me the most, although they will have to be realistic given the resources at hand. Making a pallet bench for the garden is a priority. I have also discovered a couple of fantastic gardening podcasts and look forward to listening to them as I travel or do the washing-up.

Remember to leave a few Jerusalem artichokes in the ground each year and you'll have a harvest winter after winter.

WEEK FOUR

We enjoyed regular snacks of kale chips over the Christmas period and the festive dinner was accompanied by home-grown roasted beetroot, potatoes, and leeks. Next year I would love to grow all the Christmas vegetables for free, which means sowing and planting parsnips, Brussels sprouts, and red cabbage. One idea I jotted down in the notebook was to turn the current leek bed into a dedicated Christmas dinner veg plot!

I hadn't got round to lifting any of the Jerusalem artichokes by the rhubarb, so I dug some up to roast on Boxing Day to add variety to the leftovers from the day before. The artichokes have been so productive that just a single plant yielded enough tubers. According to the forecast, a long spell of cold weather is imminent, and I can see myself making batches of warming Jerusalem artichoke soup throughout January and February.

JANUARY AND FEBRUARY

JANUARY

In early January, I spent a crisp winter's morning building a perennial herb bed, 1m (3ft) wide by 1.6m (5ft) long, out of stones and old bricks. I lifted all the perennial herb cuttings I'd heeled in, and now have mint, lemon balm, sage, rosemary, chives, thyme, marjoram, and lovage planted out. I am so excited at the thought of getting creative with them in the kitchen later this year. There is enough space for a few more herbs, so I will either source some or plant some annual herbs, such as parsley, in the gaps.

My wish for a gardening magazine subscription was granted, and by the end of the month I had gathered a small selection of free seeds – radish, carrots, peas, lettuce, and summer cabbage. Radishes taste good, but I have seeds left over from last year, and I also saved lots of pea seeds from my own crop, so I decided to set aside these two seed packets. I'll keep them in a container with others destined for the seed swap and see what's on offer.

I also prepared some ground for the soft-fruit cuttings by removing a small area of turf and spreading a shallow layer of compost over the top. I then pushed the cuttings into the ground and mulched with woodchip to help retain moisture and prevent weeds from taking over in spring.

Half of the leeks are still in the ground waiting to be lifted, and I'm picking kale on a regular basis – a reliable supply of greens is a necessity in the depths of winter. Increasing production is my key goal for this coming year, given I'll have compost from my second bin ready to use in spring, and the third bin that I built last August is almost full. I have lined and filled six more tyres with topsoil and compost, and have also created another raised bed from pallet wood. I plan to fill it using the *hügelkultur* method (see p39) in spring.

It is always a long wait for when it comes to harvesting leeks but it is totally worth it and they are such an underrated winter staple.

There's nothing better than seeing the first shoots of the growing season – here rhubarb – after a long, dark winter.

FEBRUARY

It may still be winter, but there are already signs of new growth in the veg beds. I removed the sticks and leaves from the pallet-collar bed where I planted the garlic when I saw it had sent up small green shoots. I also noticed new shoots appearing from the rhubarb divisions I planted last autumn, but I won't harvest many stems this year, so the crown can put on strong growth. I'm looking forward to picking the flowering shoots from my kale because I'm getting a bit bored with cooking and eating the leaves!

In mid-February I sowed a batch of tomato and pepper seeds I'd sourced and saved from shop-bought fruit last year. I plan to grow the seedlings in tin cans and take them to the local plant swap in a couple of months. Somehow I managed to eat the whole potato harvest (including the ones I'd set aside to grow) and have been on the lookout for seed potatoes. Luckily, I managed to get a few in return for some leeks, but I'm also going to save potato peelings with eyes in the hope that they will sprout. I'm finding the business of swapping and bartering much easier as time passes, and it will just get better because I'll soon have a greater range of seeds, plants, and cuttings to offer.

Many neighbours have burned pruned branches from shrubs and trees over winter and I have collected a respectable volume of wood ash from their bonfires. I'll mulch the soft fruit bushes and add some to the compost bin. I also gathered pieces of charcoal from the ashes, which I crushed and added to the third compost bin when I turned it.

The final winter task was topping up the woodchip on the paths around the tyre planters and some of the raised beds – I didn't want to slip on mud after heavy rainfall. The soil in the beds is clear and ready for growing but it hasn't yet warmed up. I must hold back from sowing because I know from experience that impatience often ends in disaster, and I can't afford to take risks, especially when it comes to growing food for free.

To distract myself, I've been looking through the ideas notebook and have started gathering things together for new projects. One of the easiest items to source was some old guttering. My neighbour had removed some from his garden shed and it was sitting in a skip – I couldn't walk past without asking if I could take it off his hands.

I HAVE BEEN SHOCKED BY HOW EASY IT IS TO FIND WORKAROUNDS AND COMPROMISES WHEN IT COMES TO GROWING FOOD FOR FREE, AND I FEEL EXTREMELY OPTIMISTIC ABOUT CONTINUING THIS CHALLENGE IN THE COMING GROWING SEASON.

INDEX

RESOURCES

USEFUL WEBSITES
Gardening sites
Grow Food Not Lawns growfood-notlawns.com
Guerilla Gardening guerillagardening.org
Last frost dates plantmaps.com

Community growing schemes
Incredible Edible incredibleedible.org.uk
Ron Finley ronfinley.com/the-ron-finley-project
Social Farms & Gardens farmgarden.org.uk/resources

Miscellaneous
Freecycle freecycle.org

INSPIRATION
Recommended YouTube Channels
- Back to Reality
- Charles Dowding
- Edible Acres
- Erica's Little Welsh Garden
- Happen Films
- James Prigioni
- Liz Zorab – Byther Farm
- MIgardener
- One Yard Revolution
- Rob Greenfield
- Self Sufficient Me
- UK Here We Grow
- The Welsh Gardener

Recommended books
- *Allotment Month by Month*, Alan Buckingham
- *Grow All You Can Eat in Three Square Feet*, DK
- *How to Grow Winter Vegetables*, Charles Dowding
- *No Dig Organic Home and Garden*, Charles Dowding and Stephanie Hafferty
- *Edible Paradise*, Vera Greutink
- *Gaia's Garden*, Toby Hemenway
- *Sepp Holzer's Permaculture*, Sepp Holzer
- *The Complete Book of Practical Self-Sufficiency*, John Seymour
- *Practical Self-Sufficiency*, Dick and James Strawbridge

ACKNOWLEDGMENTS

Author's acknowledgments
Writing this book has been tough, but also extremely rewarding, and none of it would have been possible without the support of my family, friends, and followers, who have been so understanding and nurturing during this process – writing a gardening book during the growing season is no easy task!

I want to thank the whole team at DK, who have helped me so much and have been instrumental in making my dream of writing this book a reality (*see opposite*); I especially want to thank my Editor, Toby, for his patience and for always ensuring I never deviated too far off track, and my Managing Editor, Stephanie, to whom I'd like to wish the very best of luck in her new ventures – I am so grateful for her kindness and the laughter she has brought me over the past couple of years. Thank you, too, to my Publisher Mary-Clare for showing faith in my ideas and for making this all possible.

One day a week, I sat in my local café, the Old Printing Office, writing this book. I truly appreciate them being so tolerant of my continued and lengthy presence – be warned, I haven't finished my weekly visits yet! Thank you too to my literary agent, Laura, for her continued and invaluable support during this process.

It is surreal to think that this is my second book, as I still haven't quite got over the excitement of publishing my first last year. I am grateful to everyone who bought this book and/or my previous one, *Veg in One Bed*, the popularity of which has entirely blown me away. My objective is to help as many people as possible grow their own food, and, for me, writing books is a massive part of that and it wouldn't be possible without you, my readers, and everyone who supports the work I do.

I can't finish without thanking my Dad, Steven, to whom I am eternally grateful for inspiring me to get into gardening and for keeping the garden in tip-top condition as I worked on this book. Your love, expertise, and support is irreplaceable.

Publisher's acknowledgments

The publisher would like to thank Amy Cox for design assistance, Steve Crozier for image retouching and colour work, Millie Andrew for proofreading, and Vanessa Bird for creating the index. We'd also like to Amy of @amyskitchengarden (Instagram) for allowing us to use her beautiful images.

Picture Credits

The publisher would like to thank the following for their kind permission to reproduce their photographs:

(Key: a-above; b-below/bottom; c-centre; f-far; l-left; r-right; t-top)

Alamy Stock Photo: Photos Horticultural / Avalon / Photoshot 167br, Michael Scheer 19, Colin Underhill 92cra, veryan dale 59, Rob Walls 58br; **Dorling Kindersley:** Peter Anderson / RHS Hampton Court Flower Show 2014 92tc, Alan Buckingham 113bl, 167bl, Mark Winwood / RHS Wisley 161tr; **Dreamstime.com:** Airborne77 64bl, Milton Cogheil 165tr, Coramueller 34bl, Jlmcloughlin 88, 99bl, Photozirka 26tl, Phana Sitti 34br, Jason Winter 67bl; **Amy's Kitchen Garden:** 21tl, 21tr; **Getty Images:** Chn Ling Do Chen Liang Dao / EyeEm 92cb; **Huw Richards:** 4tr, 5tr, 6bl, 6br, 7bc, 7br, 11tr, 11br, 17bl, 18, 22, 23, 25cl, 26bl, 26br, 28bl, 31 (a, b, d, e), 32, 33, 34tl, 34tr, 36bl, 37, 38, 39, 40, 41, 44, 47 (a, c, d, e), 48, 49tr, 55tr, 58t (a, b), 60-61, 64br, 65bl, 66bl, 67br, 70, 71, 73, 74, 75 (4, 5), 82, 83, 85 (tl, tr & br), 86, 91 (tr & br), 95, 96, 97, 98, 99tl, 100-101, 102br, 103bl, 104-105b, 108-109 (1, 2, 3, 4), 110-111, 115, 119br, 121, 122bc, 125, 130, 131cl, 132br, 135, 136br, 138br, 139, 140, 141, 142r, 144 (a, b), 146, 149b, 151 (t & br), 153 (t & br), 154, 155, 157b, 161b, 162-163, 167 (f), 174tl, 177-178, 180-182, 186, 188-189, 190tr, 192-198, 208-209, 211, 213-215.

All other images © Dorling Kindersley
For further information see: **www.dkimages.com**

Editor Toby Mann
Senior designers Glenda Fisher, Barbara Zuniga
Senior editors Anna Kruger, Dawn Titmus
Designer Amy Child
Editorial assistant Millie Andrew
Senior jacket designer Nicola Powling
Jacket coordinator Lucy Philpott
Pre-production producer David Almond
Senior producer Stephanie McConnell
Managing editor Stephanie Farrow
Managing art editor Christine Keilty
Art director Maxine Pedliham
Publishing director Mary-Clare Jerram

Photograpy Jason Ingram, Huw Richards
Illustrator Amy Cox

First published in Great Britain in 2020 by
Dorling Kindersley Limited
DK, One Embassy Gardens, 8 Viaduct Gardens,
London, SW11 7BW

A CIP catalogue record for this book
is available from the British Library.
ISBN: 978-0-2414-1199-5

Printed and bound in Slovakia

A WORLD OF IDEAS:
SEE ALL THERE IS TO KNOW

www.dk.com

ABOUT THE AUTHOR

In 1999, Huw Richards moved from Yorkshire to mid-west Wales with his parents, who were after "the good life". They settled down in the foothills of the Cambrian Mountains, buying an 11-acre smallholding, which they transformed into an abundant, nature-rich environment.

At three years old, Huw was helping his parents in the vegetable garden. Aged twelve, he created his own **YouTube channel, Huw Richards – Grow Food Organically**, about vegetable gardening. He now has over 170,000 subscribers, and his videos have collectively been viewed over 30 million times.

Since finishing school in 2017, Huw has set out to help people reconnect with the food they eat and to empower them to grow their own food, be it on a windowsill, in a garden, or in a field. Huw also does a lot of work to inspire the next generation of growers, and hopes that every school in the UK can embrace gardening as a facilitator of learning and knowledge.

Huw has been featured in *The Times*, *The Guardian*, and on BBC News, and writes a column for *Grow Your Own* magazine. He has also appeared live on BBC's *The One Show*. In 2019, Huw released his first book *Veg in One Bed*, also published by DK.

If he isn't in the garden or at his computer, he will most likely be playing tennis or squash.

Huw can also be found on **Instagram at @huws_nursery**.